IMPERIAL WEATHER

INTERSECTIONS: HISTORIES OF ENVIRONMENT, SCIENCE, AND TECHNOLOGY IN THE ANTHROPOCENE

SARAH ELKIND AND FINN ARNE JØRGENSEN,
EDITORS

IMPERIAL WEATHER

METEOROLOGY, SCIENCE, AND THE ENVIRONMENT IN COLONIAL MALAYA

FIONA WILLIAMSON

UNIVERSITY OF PITTSBURGH PRESS

Published by the University of Pittsburgh Press, Pittsburgh, Pa., 15260

Manufactured in the United States of America
Printed on acid-free paper
10 9 8 7 6 5 4 3 2 1

Cataloging-in-Publication data is available from the Library of Congress

ISBN 13: 978-0-8229-4842-1
ISBN 10: 0-8229-4842-7

Cover art: J. E. Taylor, "Chinese junks in the rain at Tanjong Katong," 1879. Watercolor. Part of *Sketches in the Malay Archipelago*, an album of watercolors and photographs made and collected by J. E. Taylor. Courtesy of the Wellcome Collection, 31904i.
Cover design: Alex Wolfe

For Pearl and Dora

CONTENTS

ACKNOWLEDGMENTS

This book has taken several years to come to fruition, developing out of research undertaken in Singapore, Malaysia, and the United Kingdom, and I am indebted to many individuals and organizations for their help along the way. The majority of this research was undertaken during the time I spent working in Singapore at the National University of Singapore (NUS) and at Singapore Management University (SMU) and I have directly benefited from grants awarded while I was at both these institutions, including from the Singapore Ministry of Education (MOE) under which I received support for two projects: the MOE's Tier 1 Grant (MOET1 19-C242-SMU-003) "Reconstructing El Niño in Singapore and Malaysia: A Multi-disciplinary Approach" and a Tier 2 Grant (MOE2018-T2-2-120), "Heat in Urban Asia: Past, Present and Future." Most significantly, I have been a multi-year recipient of the Lee Kong Chian Fellowship Award at SMU. These grants have meant that, over the past few years, I have been lucky enough to benefit from the help of many curious research assistants who have worked with me on ad hoc tasks, some of whose archival searches have resulted in examples used in this book. Special thanks go to Joshua Goh Ngee Chae who has tirelessly searched newspapers and old journals for me on a variety of weather-related subjects.

The grants also enabled me to make extended visits to British archives in London and in Cambridge. I would like to thank the wonderful staff at all the archives I have worked in, including the National Archives of Singapore and the National Archives of Malaysia, The National Archives (UK), the British Library, the Royal Society Archives and the Wellcome Library in London, and the Whipple and University Libraries in Cambridge for their help during these visits. Thanks are also due to Paul Warde who enabled me to finish my manuscript while basking in the intellectual vibrancy of the Centre for History and Economics at the Faculty of History, University of Cambridge, during the summer of 2023.

I would like to thank all the people I have had discussions with, or read, over the past years on the subject of the history of meteorology, environmental history, and the history of science. This includes the network of people involved in the International Commission for the History of Meteorology, who have provided a constant source of motivation, and all the scholars looking at the history of weather from different field perspectives, including but not limited to Aitor Anduaga, James R. Fleming, Vladimir Jankovic, Martin Mahony, and Ruth Morgan. The eminent Southeast Asian historians Timothy P. Barnard and James F. Warren have also provided considerable support, inspiration, and knowledge over the past years. I would like to take this opportunity to extend a special thanks to Rob Allan and Clive Wilkinson who first set me thinking about weather and the history of meteorology when I was a PhD student, a subject at that time very far from my chosen studies. Most recently, I am also grateful to the anonymous reviewers of this book for their generous comments, which have helped me adjust and refine the final narrative.

Finally, I would like to mention the ongoing support of my mother, Pearl Crossfield, who instilled in me from an early age the value of history and who has supported my long educational journey throughout my childhood and adulthood. I couldn't have done this without you.

A NOTE ON SOURCES

A disproportionate percentage of the surviving records for this period and region were left by government institutions—from central departments to disparate district offices or municipalities, to the correspondence and recollections of colonial officials and expatriates. The sources are biased toward the structures of state and bureaucracy and to European and British people living in colonial Malaya. Such records paint a lopsided view of life in the colony, but in this book I focus on the official narrative, not because of the availability of sources but because scientific history is one side of the official narrative that has been largely forgotten.

The primary materials of historical Singapore and Malaysia are spread across the world. The National Archives of Singapore (NAS) hold materials gathered from across the libraries and archives of the world that relate to that nation—often official correspondence and government documentation from the English East India Company through to the British government. These archives also host a remarkable oral history collection, a project that began in the 1980s and which thus holds narratives that often pre-date the Second World War. The National Archives of Malaysia hold information at the district office level, and these records are often far more interesting than the official or annual reports, detailing as they do the nuances of, say, a flooding event or a district officer's thoughts on the causes of environmental degradation in particular areas. The National Library of Singapore contains an array of texts relating to civic affairs, often in the form of annual reports, and it also hosts the incredible online repository that is NewspaperSG, a fully searchable platform of all published newspapers from the 1840s through to the modern day. I should also mention the online Biodiversity Heritage Library's dedicated Singapore collection, which includes documents contributed by the National Library of Singapore and other local institutions including the Singapore Library of Botany and Horticulture. The Singapore collection contains a wealth of information about the region's environment.

Because of the region's colonial history, many original records reside in the archives and libraries of the United Kingdom. Particularly rich sources of information can be found at The National Archives UK (TNA) in Kew, London, which holds Colonial Office (CO) documentation, and War Office, Air Ministry, and Foreign and Commonwealth Office records. These departments all had a hand in the development of meteorological science and the various means to manage the weather in British Malaya. The Cambridge University Library (CUL) is home to the Royal Commonwealth Society (RCS) and the British Association of Malaysia and Singapore archives, which contain a wealth of personal letters and narratives stemming from the experiences of British expatriates who lived in Malaya during the late nineteenth and early twentieth centuries. The Cambridge University Library also houses the Royal Greenwich Observatory (RGO) archive, and correspondence that details Greenwich Observatory's dealings with colonial meteorological and astronomical observatories.

ACRONYMS USED THROUGHOUT THIS BOOK

BAAS — British Association for the Advancement of Science

BL — British Library

BLN — British Library Newspapers

CO — Colonial Office, The National Archives, UK

CUL — Cambridge University Library, Cambridge, UK

EEIC — English East India Company, British Library, UK

FS — Federal Secretariat, National Archives of Malaysia, Kuala Lumpur

FMS — Federated Malay States

IMO — International Meteorological Organisation

IOW — Indian Ocean World

IOR — India Office Records, British Library, UK

JMBRAS — Journal of the Malaysian Branch of the Royal Asiatic Society

MAC — Malaysia Air Charter

MOE — Singapore Ministry of Education

MMD — Malaysian Meteorological Department

MMS — Malayan Meteorological Service

MPU — Malayan Planning Unit

NAM — National Archives of Malaysia

NAS — National Archives of Singapore

NUS — National University of Singapore

PCMO — Principal Chief Medical Officer

PDSI — Palmer Drought Severity Index

PUB — Public Utilities Board

RAAF — Royal Australian Air Force

RCS — Royal Commonwealth Society

RGO — Royal Greenwich Observatory

RS — Royal Society

SSGG — Straits Settlements Government Gazette

TNA — The National Archives UK

WMO — World Meteorological Organization

IMPERIAL WEATHER

INTRODUCTION

THE DAWN OF IMPERIAL WEATHER

> Singapore is unusually favoured in the matter of climate. Situated close to the Equator, it nevertheless enjoys climatic advantages not shared by other places of the same latitude. The abundant rainfall tempers the fierce heat of the Tropics; and violent storms are unknown. There is no change of seasons; the island boasts an eternal summer, and is clothed with a perennial green. . . . Singapore should be one of the healthiest places in tropical latitudes and is so for Europeans . . . the effect of the eternal summer is somewhat relaxing and enervating to those who have come from temperate regions.[1]

This was how George Murray Reith—Presbyterian minister, historian, and author of the first comprehensive tourist guide to Singapore—described the island's climate. His view of the tropics was one commonly found in Western popular culture and scientific writing about tropical environments and, especially, nineteenth-century colonial Malaya.[2] Despite Reith's long experience in Singapore, his description was full of romanticisms and scientific inaccuracies, an idealized view of the tropical island. Singapore is in fact situated just one degree north of the equator so, while it does experience year-round high levels of heat, humidity, and rainfall, there are seasonal patterns dictated by the monsoon and also extreme weathers such as the Sumatra Squalls of June to November characterized by pre-dawn winds and heavy rains. There are also heat waves, regular floods and droughts, the latter phenomena at their worst when affected by the periodic cycles of the El Niño Southern Oscillation and the Indian Ocean Dipole. Of course, Reith would not have been privy to such science, but the vagaries of the local weather could have been understood experientially.

For the inhabitants of the Malayan Peninsula, the climate was everything. It contributed to ways in which homes and clothes were designed, the times of day at which work or play occurred, the types of flora and fauna and the foods they produced, including "all the rich-flavoured, coloured fruits . . . whose generous juices are drawn from the moist and heated earth, and whose flavours are the imprisoned rays of the fierce sun."[3] The climate also dictated the ritual year. The arrival of *phinisi*, for example, sailing vessels manned by the *Bugis* of the eastern Indonesian archipelago were an annual highlight. Coming to Singaporean shores before the northeast monsoon of late November each year, the Bugis traders would erect a colorful market of cloth, sarongs, and tropical birds on the beach.[4]

For the people who migrated to take advantage of the economic opportunities offered by the burgeoning colony, the endless sun, heat, and seemingly incessant humidity were not always as romantic as Reith described. Despite the brilliant colors and flavors, the constant heat and humidity sapped the strength of Western travelers and settlers. Writing in the 1830s, for instance, Captain Sherard Osborn of the British Royal Navy, notwithstanding his extensive experience in India and Southeast Asia, still noted how, at "the residences of the wealthy European merchants; all was as dreamy, sleepy, quiet, and picturesque as anyone could desire, and I am bound to say as hot, for there the bright equatorial sun was pouring down without shadow or breeze to take off its effects. The sepoy sentry seemed to be frizzling in his leathern shako and hideous regimentals, and the sensation I felt on regarding his scarlet coat was that he might at any moment burst into flames."[5] "It is hot—so hot!" also protested travel writer Isabella Bird in the 1870s, "one does long to take off one's flesh and sit in one's bones."[6] Australian miner turned planter Ambrose B. Rathbone, writing in the 1890s, would have agreed, groaning how "the climate is a moist, depressing heat which slowly but steadily undermines the European constitution."[7] Such are the Western ideas of the tropical climate with which historians are by now familiar.

Victor R. Savage wrote in the 1980s on the subject of outsiders' perceptions of Southeast Asia, views that were socially constructed and frequently based on the strangeness of the climate and environment.[8] There has since been a proliferation of literature on "tropicality," shining a light on colonial imaginaries and artistic representations of the tropical world.[9] Historians of medicine have shown how these perceptions went on to color Western views on race, the body, and health.[10] David Arnold's study of colonial India, for example, revealed starkly how British colonials conceptualized a distinct "tropical disease environment" that shared an intimate relationship with racist stereotypes of Indian people and created colonial anxieties about the possible effects of the tropical environment and climate on the European body.[11] As Edward Said observed in his critical 1978 text *Orientalism*, racism was a fundamental component of a binary conception of temperate versus tropical that developed only in the West—in other words, the view reflected how Europeans believed the world was divided between the civilized and the other.

Reith was twenty-six when he was posted to the Straits Settlements, and during his lifetime, covering approximately eighty years from the 1860s into the 1940s, he witnesssed a confluence of ideas, where concepts such as tropicality and climatic determinism coexisted with new mathematical and statistical frameworks for understanding natural laws, the weather, and the atmosphere. Although many contradictions endured, especially in popular literature and imagery, scientists were beginning to nullify stereotypes of tropicality. This was an incredible time for the weather sciences. The focus in meteorology shifted from individual, local, and largely empirical studies undertaken by individuals and enthusiasts to international, visual, practical, and theoretical research performed by trained specialists at scientific institutions. It was a time of standardizing methods and instrumentation, large-scale data gathering, synoptic forecasting, and technological innovations such as radio and radiosonde, which were critical to advancing weather knowledge.[12] A key contention in this book therefore is to consider this alternative side of tropical thinking rather than to tread the already well-worn ground of the many prolific and influential historians who have written on tropicality in culture and in health.

BRITISH MALAYA AS A COUNTERPOINT IN THE HISTORY OF COLONIAL SCIENCE

Although the history of the development of the meteorological sciences in Britain has been well covered from many angles from the social and cultural to the scientific, this history has rarely been linked to Britain's tropical colonies, especially those in Southeast Asia.[13] By placing colonial Malaya at the heart of this story, in this book I intend to fill a gap in the historiography of meteorological science and to bring a history of science perspective to other areas of colonial development in the region, such as environmental management. This neglect, in some part at least, reflects the lack of attention paid by contemporaries to science—or indeed weather—in the region that has led to a paucity of records on these subjects.

During the nineteenth century, organizations and institutions including the Greenwich Royal Observatory, Kew Observatory, the Royal Society, and the British Association for the Advancement of Science (BAAS) were enthusiastic participants in the gathering of global weather information. At the very first meeting of the BAAS, for instance, members requested a report on the current state of meteorological science as a precursor to including this nascent field within their future work.[14] The BAAS and the Royal Society drove a new scheme shortly thereafter to establish a series of observatories across the British Empire, including in Madras (1845) and Mauritius (1851), at which magnetic and atmospheric conditions could be studied.[15] Nonetheless, meteorology was not the prime concern of their subcommittees or the observatories. The science was subsumed under broader themes, which included the different disciplines of the mathematical sciences, light, heat, and sound. Meteorological studies were subservient to magnetic or astronomical research or were undertaken to provide a time and storm warning service for shipping rather than for meteorological research per se.[16] Certainly, the needs of the shipping industry were paramount to the British Empire, and thus the institutions most directly concerned with meteorology, notably the British Meteorological Office and its colonial counterparts, were set up with exactly this industry in mind, both at home and abroad.[17] Nineteenth-century imperial meteorology was concerned with forecasting weather events such as typhoons at sea that could affect trade or the monsoon that could affect agricultural and economic productivity on land.[18] Therefore, although the Straits Settlements incorporated the three major ports of George Town, Malacca, and Singapore, there were no cyclonic systems in the immediate vicinity, and explorers on early incursions into terrestrial Malaya suggested that the area did not suffer any extreme weathers. At the same time, it has been suggested by both contemporaries and historians that science in Malaya was behind that of other countries, as it was often left to individuals to champion rather than being a dedicated government or institution-sponsored effort.[19]

The claim of some historians that science and empire developed together has applicability to the weather sciences, but in Malaya, the perceived lack of extremes pushed the development of that science the other way. The establishment of meteorological networks and infrastructures across imperial geographies, the standardization of observational practice in ways specific to imperial and Western thinking, the use of meteorology in enhancing imperial productivity were all core characteristics of how this story played out in many other colonies.[20] But, the case of Malaya competes with the notion that British ways of doing and thinking were applied evenly across colonized regions and the concept of a homogenous imperial project is far too problematic to apply here.[21] Some of the authors of the latest work on the transfer of scientific knowledge suggest strongly that the colonial effort was neither linear nor uniform and that there was no single scientific colonial discourse or practice.[22] I am inspired by this line of thinking. In this book I therefore hold a question mark over the idea of science as a colonial tool and probe the equation of science and imperial control.

To establish an overseas settlement that would have commercial value, it was necessary for contemporaries to understand whether the climate would be suitable for long-term European habitation—that is, would it be conducive to establishing commercial crops and healthy living? Some basic meteorological facts, such as annual averages of temperature and rainfall, were needed to determine the answers to these questions, and weather records made during exploratory missions or at the point of settlement were used to justify why an island or a harbor might be worth settling. Nonetheless, the Straits Settlements—unlike other British colonies including Hong Kong, Australia, or India—did not adopt a specialized observatory or institution until the 1920s. They also resisted much of the regulation and standardization pushed by bodies including the BAAS or the International Meteorological Organisation (IMO), despite the "calls of colonialism and modern science to know and to categorize, and in so doing, control environments."[23]

In meteorological historiography, it has also often been assumed that the observational networks linking imperial hubs and outposts were reflective of, and constituent to, imperial control. The observation stations, observatories, telegraphic lines, and flows of information collectively created a web that distinguished imperial lands and connected disparate areas in an imperial web. This network was reflective of which parts of the empire held the most value—stations emerged first in urban settlements, plantations, and botanic gardens—but although their locations mimicked the geography of power, in this book I also argue that the network did not exert power in and of itself. The network generated data, but the data's usefulness was limited by problems of accuracy, use, and scope. Arguably, these networks were just as reflective of the failures in the colonial system of control and surveillance as of their strengths. Either stations did not exist in areas that were too remote, or, just as Mark Frank argues in relation to Tibetan observational stations, many of the observational stations of the Malayan interior operated semi-independently, with few checks, being reliant entirely on the ability or the dedication of the staff who were present.[24]

Although the Malayan network could have been critical in providing the information for developing our understanding of the atmosphere as a global dynamic system, in the nineteenth century at least it was subject to human error, the information generated was full of chronological gaps, and the results were not applied to contemporary scientific studies of atmospheric weather conditions. The tropical weather remained by all accounts an imagined space in colonial science, and the Malayan example thus counters suggestions that meteorology was an organized and structured part of imperial state apparatus. The situation was perhaps more akin to what Martin Mahony has described as "faltering internationalism" where local circumstances and ambitions dictated the pace and scale of investment in knowledge making.[25] Thus, in this book I argue that colonial meteorology was less about purposeful science and the imperial project and more about trial, error, and accident.

RETHINKING COLONIALISM AND SCIENCE THROUGH THE WEATHER

The arguments laid out above position weather as a lens into colonial science and imperial functionality in the tropics. With this view I will show that the British government had a lack of interest in shaping Malayan meteorology and, although colonial officials dictated the key means of managing weather's extremes, from flood mitigation to drought protection, their role was largely reactive and, on occasion, inadequate. Often it was left to individuals or private enterprise to champion a cause. Following this line of reasoning, I provide a deep exploration of colonial era actors' attempts to know, manage, and control the previously unknown or the uncontrollable through information gathering, institutionalization, and infrastructural change from the standpoint of governmental, scientific, and technological responses to local weathers. I exclude the history of indigenous weather knowledge, unless this directly informed colonial modus operandi in this field.[26] This is not to downplay the importance of local knowledge or practices but to demonstrate that colonial science—in Malaya at least—was largely restricted and, although the day-to-day operations of meteorology relied almost exclusively on a local workforce, indigenous or traditional knowledge did not inform Western styles of meteorological science or the technological and engineered responses to extreme weather that were being practiced. This is markedly different from the experiences portrayed in some recent texts on the relationship of colonialism and local knowledge, in environmental histories of jungle commodities or bioprospecting, for example, or in the exchanges on climate with indigenous communities that have been described by Deborah Coen, Philipp Lehmann, or Thomas Simpson.[27] In Malaya, this dialogue was largely absent in the world of colonial weather science or environmental control.

In this argument I draw from the history of science—especially the history of meteorology—to think around the how and the why of weather knowledge within the tropical Malayan context. There have been recent studies undertaken on the environmental history of Malaya, but there have been no studies that foreground weather as a component in Singaporean or Malaysian history nor any that merged the history of colonial environments with the scientific history of weather and its extremes.[28] My premise is that these fields are not distinct and ought to be considered in their relationship to one another. Critically, I consider the weather as a catalyst for the development of science and land resources management, lending agency to the atmosphere in the shaping of colonial science and governance, showing how local circumstances often triumphed over global knowledge in the execution of science. Doing so foregrounds weather within a framework of colonial interests, capacities, and capabilities, especially the state of science and technology, the desire to have a productive economy, and the realities of the environment.

Tropical weathers and landscapes presented extremes and uncertainties. These necessitated the acquisition of new knowledge to maintain settlers' quality of life and their economic resilience by controlling, or trying to control, nature. Nature dominated every aspect of life in the tropics, and it was almost impossible to tie it down. As the local writer Walter Makepeace was to note in the early twentieth century simply of the realities of indoor white-collar work: "the climatic and geographical conditions of Singapore affect the personnel of the firms that carry on business here (since practically every business man has to go on frequent leave) and the length of the life of the firms."[29] His viewpoint spoke only to the experience of the colonial European of course, but the challenges of creating a livable and workable environment in a tropical atmosphere often pushed the colonial government beyond its capacities. Solutions were frequently enacted at the expense of nature, which, ironically, often made managing the effects of the weather more problematic in the longer term. The deforestation needed to create new plantations, for instance, caused problems of soil degradation and erosion that, in turn, exacerbated the impacts of floods and droughts. This also began a long process of altering the indigenous landscape in ways that were never recovered.[30]

TEMPORALITY AND CONTEXT IN COLONIAL MALAYA

The focus here is on the multicultural ports and peripheries of Singapore and to a lesser extent George Town, Penang, the second-largest town of the Straits Settlements, although I also discuss Malaya more broadly. The temporal scope of the study is the nineteenth and the early twentieth century, during which time both the English East India Company (EEIC) and the British government had varying degrees of political and economic control over the Straits Settlements, the Federated Malay States, and the Unfederated Malay States. It was a time of sweeping social, economic, political, environmental, urban, climatic, technological, and scientific change.

The peninsula's strategic location between India and China enticed the British to challenge Dutch trade dominance in the East Indies by their settling Penang Island, which the British called Prince of Wales Island off the northwestern peninsula, and Singapore Island to the south. The ports of George Town, Penang, and Singapore were established through complex political negotiations with local rulers and force of personality. From the 1770s onward, the northeastern border region of Kedah was in turmoil because of a dynastic dispute and successive incursions from Bugis and Siamese factions. The sultan of Kedah, Sultan Muhammad Jiwa, sought military aid from the

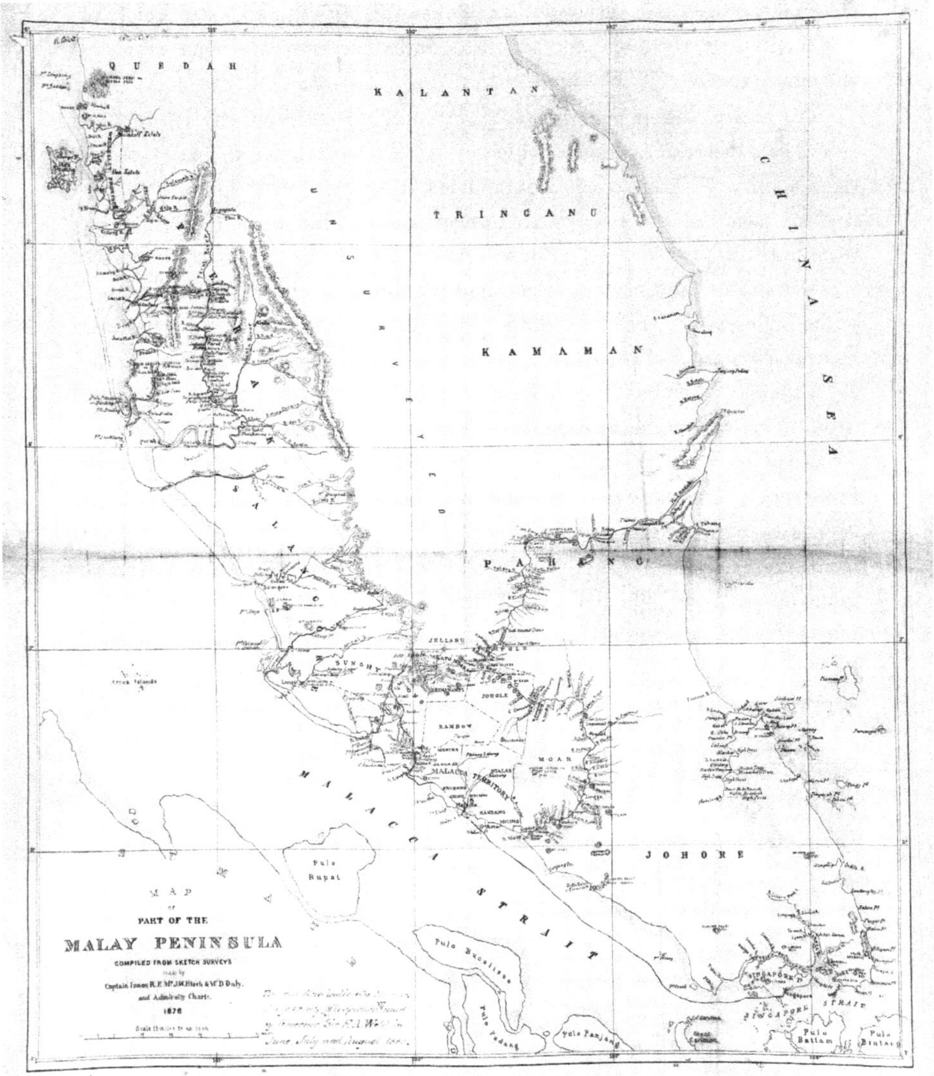

FIG. 1. "Malay Peninsula," 1876. MPG1-1023, TNA.

EEIC in exchange for a monopoly on tin, pepper, and ivory, a practice that continued under his son Sultan Abdullah Mukarram Shah. Captain Francis Light of the EEIC saw opportunity in the situation and his eventual cessation of the site of George Town on Penang Island in 1786 began the direct intervention of the British into Malay territory that extended to the whole of Penang Island and Province Wellesley on the mainland.[31] The settlement at Singapore Island was established through a negotiation between Stamford

Raffles, then lieutenant-governor of Bencoolen, Major William Farquhar (who would become the settlement's first Resident), the Temenggong Abdul-Rahman of Johore and the sultan of Johor-Riau and Singapore, Hussein Mua'zzam Shah ibni Mahmud Shah Alam, between 1819 and 1824.[32] The settlement was originally governed from Bencoolen in Java from 1819 but it was placed under the control of the governor-general in India, via EEIC administration from 1826. It became the administrative capital of the Straits Settlements (which also included Malacca and Dindings) in 1826.

From the very beginning, the placid climate, along with the strategic position of the settlements, would be a boon. Exploitation of the peninsula's natural resources followed quickly, engendering the development of the major nineteenth-century trading entrepôts for the region—George Town and Singapore—and the transformation of land for commercial purposes. After the EEIC failed in the 1860s, the Straits Settlements became a Crown Colony and the opening of the Suez Canal in 1869, combined with the transition from sail to steam, created the conditions for trade expansion. The British extended their influence further into the mainland after the 1874 Pangkor Treaty began the British Residency system in Perak, leading the way to formalizing the Federated Malay States (comprising the agricultural states of Perak, Selangor, Negri Sembilan, and Pahang) in 1895. This opened these agricultural states to free trade and allowed the British to develop and control their commercial plantation agriculture. Early commercial enterprises exporting tin, gutta-percha, coffee, and pineapple were rapidly overtaken by the commercial success of rubber in the early 1900s.

All of this enacted a heavy toll on the environment, however. Mining and agriculture, urbanization, and infrastructure, deforestation, land-use change, alterations to watercourses and ecosystems, all meant that human and animal resilience in the face of extreme tropical weather declined. Local governments, planners, and engineers responded by trying to improve their knowledge of the weather and its dynamic interaction with the local environment, to create effective solutions so as to manage the most extreme of the weather's impacts on the towns and agricultural districts. The pace of man-induced environmental change was clear to contemporaries. Some began to question man's relationship with nature, especially regarding deforestation. They saw virgin forest give way to commercial monocrop plantations, soil erosion on denuded hill slopes, and rapid urban expansion. They recognized that barren hills and riverbanks led to an increase in the frequency and severity of land slips and floods and wondered, drawing from current scientific thought, whether the loss of natural habitat was causing changes to localized rainfall. They even theorized on whether urbanization was causing temperatures to rise.

Between the 1870s and the 1920s, anxieties increased as the El Niño

Southern Oscillation (ENSO) brought unusually extreme weathers to the British Empire across all of Asia. The ENSO is a cyclical atmospheric phenomenon. It has three main phases—neutral, positive, and negative, the latter two (known as El Niño and La Niña) result in low and high rainfall respectively. In Asia, these phases can induce devastating social impacts from droughts or floods.[33] Some phases can be stronger than others, such as the exceptional El Niño of 1877 or that of 1911. ENSO's interaction with the regional Indian Ocean Dipole (IOD), a three-phase climate oscillation that affects the temperature of the Indian Ocean, can also exacerbate the effects of the ENSO. The ENSO phenomenon was not identified formally until 1924 and the IOD not until the 1990s, so during the time covered in this study, observers identified patterns and speculated on whether human-induced environmental degradation or other phenomena such as sunspots caused this periodic extreme weather.

Ramachandra Guha has argued that the impacts of colonialism on the Indian environment, environmental destruction and climatic instability, attacked traditional sustainable and self-sufficient ways of living, leading to social dislocation and enhanced vulnerability. So too was the situation in Malaya.[34] Land-use change, imported engineering, population increases (especially of new urban inhabitants), and the rise of the technocrat divorced people from traditional ways of living with nature and altered the environment in ways that benefited the colonial class rather than the rest of society.

CLIMATE AND COLONIAL SCIENCE

Reliant as they were on a stable environment to maximize agricultural output and minimize outlay from damages, the colonial government officials were incentivized to increase their spending on mitigation measures but not to increase their research into understanding the atmosphere. Historians of science will be familiar with the concept of how scientific knowledge develops in site-specific places with site-specific needs.[35] Studies of the field, the lab, or the institution as the site of knowledge production have much applicability to the Malayan case as, although the colony was not a lab or an institution, it is plainly apparent that local weather and circumstances forced the balance of need, over knowledge, in managing the weather.[36] The tropical situation therefore offers new conceptual opportunities for us to explore how intemperate climates forced new ways and scales of practicing science and environmental management.[37]

However, this is not a story of progress toward a scientific apotheosis. The story of Malayan weather science and weather management is a story full of gaps, of failures, and of disinterest. Knowledge of tropical weather and its extremes remained largely undeveloped, despite advances in meteorological and climate sciences in the global North. In the first half of the nineteenth century, the sciences of tropical weather and of regional weather systems were a blank. Although there were efforts made to understand general information as to the typical weather of new settlements, such as the mean temperature and rainfall, from a scientific perspective these were limited to observations over short periods of time from occasional station points.[38] Observational and theoretical work was undertaken in the field of cyclonic storms of the Indian Ocean and the China Seas, mainly supported by mercantile and military interests, but the understanding of tropical monsoon systems or simply everyday weather in Malaya was largely absent.[39] Over the course of the century, theoretical and empirical knowledge about global climates shifted from thinking about climate as stable and localized to knowing it as a changeable phenomenon. These understandings developed largely outside of any empirical research in Malaya, however; the chief theory to impact on scientific thinking and practice was a shift in the knowledge of the scales at which weather, or climate, operated.[40]

In the early 1800s, colonial climates were measured at local scales. Military officers of the EEIC might record the weather in port or, occasionally, at a civic building or hospital. Other interested observers made single-site records within close distance to developing towns and their peripheries.[41] These records were rarely compiled for more than a year or so, in part because of the transient nature of the observers or their instruments—perhaps they were an EEIC officer who was posted to a new location—but also because it was supposed that there was no scientific reason to do so. The climate of the tropics was thought to be stable all year-round, year in, year out, and the peaks and troughs caused by the ENSO and IOD were not picked up on until much later, through a combination of lived experience and the availability of longer-term records across wider areas. As the Malayan interior was cleared for plantation, the number of station points built to make observations were extended outside of the urban settlements of Singapore, George Town, and Malacca. Patterns identified in the records were connected to contemporary theories coming out of Britain and Europe—Luke Howard's or Emile Renou's work on city heat in London and Paris respectively, for example—and it was also suggested that rainfall may have declined in areas where forests were cleared, an ancient theory reinvigorated during the 1800s and referred to as desiccation.[42]

As the century progressed, observers also began to develop a keener sense of climatic teleconnections across regions rather than just across localities or districts. A major drought in 1877–1878, for example, made meteorologists across Britain's Asian Empire—such as Henry Blanford, head of the newly established Indian Meteorological Department (IMD), and Charles Todd, the government astronomer in Australia—only too aware of the possibility that climate might operate on regional or even global scales. Blanford, using synchronous instrumental weather observations gathered from a network of newly established regional stations, deduced that the "condition of excessive pressure prevailed over not only the Indo-Malayan region and Eastern Australia, but also the greater part if not whole of Asia, probably the whole of Australia and the South Indian Ocean." Todd similarly concluded that "there can be little or no doubt that severe droughts occur as a rule simultaneously," referring to reports from India, Singapore, and Batavia.[43] Such remarks were only possible because of the expansion of the observatory and station network across Asia and of collaboration between different nations in and around Asia from India, to China, to Hong Kong, the Philippines, and Japan, on matters of weather reporting and forecasting.[44] In this way, the scales of weather and its impacts metamorphosed from local weathers to regional climates and the enduring legacy of climate as a stable entity was transformed.

This new knowledge—especially that stable climates had natural or human-induced fluctuations—fed into the discussion of hydraulic management. Heavy rains or flash floods were a regular feature of tropical living, but extensive forest clearances, clean weeding practices on plantations, and mono-cropping laid the groundwork for intensifying soil erosion, runoff, and flooding. George Town, for example, was bounded by steep hills to the west, and it was clear to everyone that development on the hillside intensified soil erosion, and thus flooding and landslips, impacting on the public health of urban inhabitants, as well as causing regular damage to the urban landscape. Repeated droughts eventually forced the local government to improve the city's system of reservoirs and piped water, especially as urban populations rose.[45] Although hydraulic systems are almost as old as humanity itself, the colonial experience imposed foreign engineering structures into tropical environments, structures that often impinged upon the flow of nature or existing indigenous systems, in part to manage the imported ways of living that differed from extant means.[46] Controlling water required huge, expensive infrastructural investments, and it has been argued that the resultant hydraulic systems enabled colonial government officials to better utilize their colony's resources, protecting and enhancing economic productivity.[47] A close study of the development of hydraulic management in Malaya, however, reveals a more nuanced account, where infrastructures routinely lagged behind need.

By the early decades of the twentieth century, the frameworks, methodologies, and technologies available for thinking about, or managing, weather had expanded. In the scientific world climate was considered a teleconnected and global experience. Evolving communications technologies from telegraph to wireless and the transition from steam to air enabled meteorologists to gather data at temporally current, geographically broad, and infrastructurally embedded scales. Weather reports and forecasts were critical to improving agricultural yields or to cultivating new areas and were essential to the nascent aviation industry.[48] Improving meteorological capacities at large scales also demanded international cooperation, agreement, and standards. But just as weather science expanded into these macroscales, urbanization and new technologies created the possibility of microclimates.

Over the 1920s and 1930s sociotechnical changes created artificial climates, which were distinct from anything known during the previous century. Whether it was an urban street heated by motor car pollutants, hard road surfaces, and increasingly concretized surroundings or an indoor space cooled by fans and later air-conditioning, modern living created new atmospheres and microclimates. The notion of indoor and outdoor space became gradually less fluid, as buildings were progressively closed off from the outside world, shuttered to keep in cooler airs generated by technical solutions rather than kept open to attract breezes. The introduction of electricity and refrigeration in combination with improved water management systems began the long and disjointed process whereby people who could afford it became divorced from their environment. During the period of this study, however, people still lived "in climates—amidst the particular fluxes of weather that they encounter in different places, visceral experiences which are interpreted through their imaginative worlds."[49] It would be a long time before the advance of technology would impact on people's lives evenly, if at all. Within the time frame of this book (and for many even longer still), money made the difference. Money determined whether you lived in cheap, housing prone to flooding, had insurance, or could afford any form of domestic cooling system. There was no clear chronological transition, and many of the same vulnerabilities that existed in early colonial societies endured. For some, modernization led to better standards of living and detachment from the climate, but for many, a reliance on wage labor and mass market commercialization lead to heightened social divisions and thus greater vulnerability to climatic shocks.

In this book I move through all these shifts in climatic experiences within the context of colonial Malaya. From the science of meteorology to weather management, from perceptions of tropical weathers to the creation of new microclimates, the weather infiltrated all aspects of life. As this story will show, studying weather science and weather management offers an alternative way into understanding colonialism and the environment.

THE FRAMEWORK OF IMPERIAL METEOROLOGY

The book is separated into three sections, each of which considers a different aspect of weather in colonial Malaya: the science of weather, changing weather, and controlling weather. Considered as a whole, they offer a long view of change over time while facilitating in-depth insight into different areas of weather studies and human-climate interaction in colonial settings. The first section comprises an overview of the state of meteorological and climate science in the colony. I explore how tropical weather was understood by contemporary men of science and colonial officials, what their chief concerns were, and relatedly, how meteorology was practiced. In chapters 1 and 2, I explore the situation of Malayan meteorology from the 1820s until just after the conclusion of the Second World War. This was a significant transitional period in the history of meteorology globally, as well as in Malaya, with new communications technologies such as telegraph and radio increasing the speed and distance across which weather observations could be transmitted, new global institutions and standards, such as the IMO or the Beaufort Scale, new practices including mathematical modeling and theories to understand atmospheric phenomena such as cyclones. Nevertheless, revealed in both chapters is a strong theme of institutional disorganization and governmental disinterest in advancing the science in Malaya, especially in the nineteenth century, despite a growing global acknowledgment of its importance. These chapters constitute the contextual foundation for the remainder of the book.

From here onward, chronology gives way to theme. In Part II I dive deeper into the areas in which colonial officials held a particular interest, especially in response to ideas popularized in the later nineteenth century about the possibility of man-made climatic change. The focus in chapter 3 is on the desiccation "debate," an idea that had been around for centuries but which drew particular attention in the 1800s as colonial activities rapidly reshaped natural landscapes and European explorers asked questions about the desertification of African and central Asian interiors. This discussion hit a peak around the 1870s in connection to the question of whether forest clearances were decreasing rainfall, a critical concern for governments supervising plantation economies just as in Malaya. Weather records became the chief means to establishing patterns in rainfall and establishing norms and deviances.[50]

The government's desire to sustain and improve economic productivity was behind most of the main thrusts in environmental conservation or meteorological investment in the nineteenth century. Protecting the riches of the forests and jungles would provide woods for the railways and infrastructure, manufacturing, building, and furniture, and the resins, gums, and oils used in all types of products from medicine to food, dyes, and perfumes.[51] As Hugh Cleghorn, the conservator of forests in India, noted in 1861, the practice of forest clearances by unscrupulous contractors, traders, and natives needed to be stopped for the "public good" because they were laying waste to "useful material" and resources "valuable to science."[52] These were all materials that colonial officials wanted to exploit themselves. By the first decades of the twentieth century, tropical forests were seen as a source of everyday materials, such was their significance in global commodities markets.[53] As early as the 1870s, it was also thought necessary to protect the forests in order to protect rainfall in a country that depended on its plantations for its wealth. These fears resulted in first conserving areas of forestland, such as Nathaniel Cantley's proposals to conserve large swathes of Singapore's jungle in the 1880s, and second, improving the network of meteorological observation stations across the country so as to enable sufficient density and breadth of records for detecting changes in rainfall.[54]

Conversely, however, in chapter 4 I show that studying temperature was conceived of very differently from studying rainfall, despite the plentiful depictions of the heat in contemporary literature (travel writing) and the concern of medical establishments for human (European) health in hot weather. Most weather records kept across the nineteenth century comprised temperature as well as rainfall measurement. Yet there was little coordinated research on tropical temperature (outside of medical literature) within the Straits Settlements. In this chapter, therefore, I explore some of the possible reasons for this incongruity but also draw attention to the irony that, while temperature research per se was not considered a subject for meteorological science, there was a limited but steadily mounting discourse on urban heat.

The discussion on urban heat in Malaya was quite clearly influenced by scientific research coming out of Europe on how urbanization increased heat but was provoked by the lived experience of urbanization. Keeping cool was a priority for Europeans unused to the tropics, so the idea that the growing towns might be worsening the situation caught contemporaries' attention. Unlike the desiccation discussion, the urban heat discussion was more of a scientific curiosity than a concern, unlike today. Urban heat may have made things less comfortable, but small temperature differences were not considered serious enough to affect human health or have a drastic impact on productivity, unless there was an exceptional heat wave. Also, at the time that increasing development, concretization, and motor cars were making urban living hotter, corresponding coping technologies were being introduced to cool things down, such as the electric fan and, later, air-conditioning. Modern science was providing the technological solutions to improve lived experiences through controlling the climate, even if it could not—yet—make it rain.

Controlling the climate is thus the theme that binds the third and final part of the book. Part III is an interrogation of the conceptualization of imperial control. Using weather—an inherently unstable and unpredictable phenomenon—means a reevaluation of the relationship between colonialism and power. The story of how, and how far, British colonial officials attempted to manage the weather's extremes—in an age where science was hoped to be the answer to living, and profiting, within precocious environments and atmospheres—is telling of this relationship.[55] In particular, Part III draws attention to two aspects to this story. First, and in keeping with the themes of the previous chapters, it points to inaction, or inability, in the colonial government's response to the weather. Second, it looks at how managing weather proved to be a reflection of, and driver for, the social divisions inherent in colonial society.

Chapter 5 is an exploration of flooding. Malaysia and Singapore have a high annual rainfall that is spread relatively evenly throughout the year, but from November to January monsoon surges can cause severe rainfall events, and during May and June there is a high incidence of flash flooding because of the phenomenon known locally as Sumatra Squalls. These squalls, when combined with a high tide, mean that low-lying coastal areas are easily flooded and, when combined with La Niña heavy rains, can cause rivers to burst their banks and mountainous areas to experience devastating landslips. Flooding was a serious problem for the expanding settlements and for the agricultural land of the colony. My argument, however, is that although basic precautions such as seawalls or embanking were established early on in British rule in new urban settlements, more extensive initiatives tended to be put in place reactively, rather than because of forward planning. This was especially the case before the 1920s and outside of the towns. Even when solutions and schemes were proposed, they were quite flagrantly aimed at protecting the town's civic and business interests over and above the town's poorer inhabitants. Any given scheme's inception and realization was frequently hampered by a lack of funds, resources, and disagreement at the top levels over what ought to be done and how. Far from the notion of control, local governments were compelled into taking measures because of the recurrence of costly disasters, a growing complaint culture among citizens, and a changing sense of expectations of the role of government in the twentieth century.

With the focusing on the converse extreme of drought, chapter 6 is an exploration of the colony's quest for sustainable and potable water, especially in the emergent towns, as it was discovered that Malaya was not always the persistently wet environment that was first assumed. Rapidly rising urban populations demanded ever more complex infrastructures and technologies to conserve and distribute water. This chapter continues the theme of omnishambles, arguing how schemes to conserve or enhance Malaya's urban water supply were reactive, were underfunded, and suffered considerable implementation delays. Notwithstanding, securing a constant and abundant water supply for harbor activities and shipping was vitally important to the economic survival of the ports of Singapore and George Town, and ensuring an adequate and healthy water supply for urban inhabitants was considered a moral imperative. Thus, larger reservoirs and ever more elaborate and expensive means of drawing water to the towns, especially Singapore, became a focus of engineers from the 1910s to the 1930s. Feats included the monumental Johor-Singapore Causeway, which in addition to its function as a land crossing, contained waterpipes linking Singapore to a reservoir at Gunong Pulai on the Malayan peninsula.[56] Over time, the improved availability of long-term weather observations, especially from the 1870s onward, invited contemporaries to investigate the seeming regularity of extreme drought periods, to have data to underpin water usage projections in line with population and industry needs, and to scientifically plan the capacities required for flood mitigation schemes.

However, the more that nature was "controlled" for some, the more others became vulnerable during periods of climatic instability. Indigenous communities such as the *Orang Laut* (sea people) survived by coexisting with the ocean's natural rhythms, or *Jiwa Laut* (spirit of the sea), and the *Orang Asli* (original people) had free rein of the jungle's incredible resources.[57] As their lifestyles and communities were gradually subsumed by the progression of infrastructure and the transformation of land from swamp and forest to plantation and town, they, like the large migrant communities of unskilled workers and laborers who had immigrated to the towns and plantations, had less recourse to traditional ways of living with water. These people were at most risk from living in poor quality housing on cheap land that was easily flooded, having limited access to fresh water supplies to manage through a drought.

Finally, in chapter 7 I consider climate control at the micro-level. Instead of looking at how unpredictability was managed, I explore how the predictable specter of heat and humidity could be managed through the introduction of cooling technologies. This process was immersed within an era of technological optimism whereby scientific expertise was considered part of the solution for resolving the problem of the climate, not through massive geoengineering schemes or climate modificationbut through small-scale changes in homes, offices, institutions, and places of entertainment. Wholly reliant on the pre-introduction of electrification, air-conditioning was symbolic of modern, clean, and healthy living. While the process of cooling Malaya was not a government-managed initiative, it was still an issue of managing and living within a tropical climate; cooling technologies were embedded within colonial-era socioeconomic structures and were largely imported from the West. Their advent and adoption presaged a new way of distinguishing between rich and poor, with cooling technology becoming a marker of wealth and better living standards. Just as the focus in chapters 5 and 6 is on socio-environmental injustices during flooding and drought, the focus of chapter 7 is on how cooling technologies were socially discriminating. Early air-conditioning was available only in spaces that charged an entry fee, such as cinemas or dance halls; domestic air-conditioning units were out of most people's price range. Thus, this chapter is an exploration of climate control as a form of sociotechnical imaginary, achieved by the few while imagined by the many, shedding light on how colonial-era aspirations to modernity were restricted to time, space, and person.[58]

Overall, in this book my aim is to speak to readers of environmental and scientific history. As such, the study is located at the intersection of both fields and may therefore seem incongruous to readers firmly situated in either one or the other. My aim, however, is to illuminate how the climate crosses disciplinary borders as well as national ones and how, when used as the means of interrogation, climate reveals the relationship of colonialism, science, and society.

PART I

IMPERIAL WEATHER

"Forecasting in the equatorial belt is a special problem," wrote the director of the Malayan Meteorological Service C. D. Stewart in 1935. "In the absence of precedents or of former experience which might have given some guidance, the methods in use are still largely experimental. It is difficult to judge whether Malayan forecasting problems are peculiar to Malaya or whether they may be regarded as typical of equatorial regions generally," he went on.[1] In 1935, the British colonial government had more than 140 years' direct experience of the climate and weather on the Malayan Peninsula. Yet, fundamental questions about the tropical atmosphere were still yet to be answered.

During the early twentieth century great strides were made in meteorological science globally, aided by new communications and detection technologies such as wireless radio transmissions and radiosondes. The advent of commercial aviation, what Robert Marc Friedman terms the rationalization of the weather, and the emergence of new schools of thought during the 1920s and 1930s meant that European, Asian, and American weather sciences had reached a space of professional and dedicated study that had been found only in pockets during the nineteenth century.[2] European nations and the United States invested in training young meteorologists during the First World War, resulting in a postwar professional body of weather experts. Most developed nations had a national weather bureau in addition to academic programs at local universities.[3]

Penned in the special report *Weather Forecasting in Malaya*, C. D. Stewart's view, however, reveals that tropical meteorology in the Malayan context was not as far advanced as weather studies elsewhere in the world. This could not be because the knowledge being generated from the new schools of thought, such as the Bergen School, was not transferred to the tropical colony; more likely it was that the specificities of the tropical atmosphere were less well understood; and, critically, many of the infrastructures and technologies that were working to advance the science in Europe were simply not extant in Malaya. As Stewart himself noted, it could take twenty-four hours for information from northern Malaya to reach the Malayan Meteorological Service Headquarters and Forecasting Centre in Singapore.[4] Thus, he argued, the precision and speed required even for efficient synoptic mapping, let alone anything else, was simply not present.

This section, comprised of two chapters, forms the first part of the book. Together, they aim to answer the incongruity that was British colonial weather science in Malaya and to to discover the reason that tropical weather research here trailed behind the rest of the world, despite Britain's political and economic interest in the region. The chapters set out the context for nascent meteorological science in colonial Malaya and provides a chronological overview of the development of weather services and the local and global catalysts for investigating tropical weather. British investment in weather science in tropical Malaya seems as something of a lost opportunity, in contrast to the weather sciences that were developing in Britain, Europe, or indeed in other parts of Asia, both within and outside of imperial contexts.

CHAPTER 1

ATMOSPHERIC STATES

IMPERIAL WEATHER AND THE COLONIAL ENTERPRISE

Meteorology is frequently considered a colonial science, said to have enabled and supported the imperial project and colony building in fundamental ways.[1] First, understanding the weather was fundamental to planning the transoceanic journeys and trading networks that were essential to any aspiring imperialist nation. In the Indian Ocean World and in the region of the China Seas, knowledge of the monsoon and the ability to predict typhoons were critical for avoiding disasters at sea and for being able to use the weather to its maximum sailing advantage.[2] Second, weather knowledge was fundamental to determining where colonists might settle and how. Would the rainfall be conducive to growing worthwhile crops? Would the climate be healthy for settlers? This local knowledge was initially acquired through observation and by working with indigenous peoples to understand seasonality and patterns.[3] Weather information was gathered, plotted, and communicated across empires and throughout scientific networks, which resulted in a reframing of space and time through a new imperial, climatic lens.[4] Meteorological focal points such as observatories and registering stations connected remote places on land and at sea, birthing an emergent reseau for imperial worlds, often in conjunction with the sciences of geomagnetism, astronomy, and seismology.[5] As the scientific world increasingly turned to statistical methodology and empiricism in the late nineteenth century, meteorology shifted in turn toward the mass collection and standardization of weather observations across this global network.[6]

Setting out the evolution of meteorological science in this teleological manner would nonetheless do this story a disservice. Frequent and significant differences in opinion in many areas of meteorological theory and practice meant dissonance rather than coherence was the norm, holding back some opportunities for progress. Just in terms of the standardization of weather statistics, for example, there was little consensus. Imperial nations created and operated their own ways of measuring, recording, and interpreting weather, such as the empirewide observatory network established to study geomagnetism by the EEIC and the BAAS in the 1830s and the German Magnetischer Verein under physicists Carl Friedrich Gauss and Wilhelm Weber. Both countries' officials used different benchmarks for recording observations at macro-scales.[7]

The very idea of an international observational standard was politically provocative at this time of intense European rivalries. Nomenclature and scale were contested, and methods and practice differed not only from country to country but from scholar to scholar.[8] Rival interpretations of storm theory caused decades of contention: the differences between the ideas of William Reid, William Redfield, James Espy and Elias Loomis are several examples. As Aitor Anduaga argues, meteorologists also found themselves in two alternative camps, preferring either Eulerian or Lagrangian approaches for explaining the dynamics of fluid and air motion (the difference contingent on the observation of atmospheric properties from fixed or moving points, the former empirical and the latter theoretical or numerically determined), but which underpinned forecasting attempts from the 1820s to the 1860s.[9]

The meteorological sciences also suffered from an uneven application of resources and interest from governments across time and space, which translated into considerable bias toward research on the global North, at the expense of the global South. Within Britain's Asian Empire, nowhere was this more apparent than in the Straits Settlements. There was limited central funding to support locally based scientific projects, which resulted in erratic interest or support for meteorology for more than a century after the British settled on the peninsula. Thus, while meteorology was used by colonial powers in many ways, its importance in Malaya waxed and waned, and as a result, the science did not play as important a role in colonization as it might otherwise have done.

Following such a line of argument, this chapter forms the first half of the chronological exposition of Malyan meteorology that makes up Part I of the book. Together these chapters foreground and provide context for understanding how British colonists conceptualized Malayan weather and practiced meteorology. The chapters argue that local circumstances were, in many cases, more significant than global knowledge, and meteorology operated at different scales and chronologies than elsewhere. Malayan meteorology was not systematized until 1929, and it lacked a clear space in colonial administrative structures. This chapter in particular suggests an interpretation of imperial meteorology that is haphazard rather than coordinated, accidental as opposed to directed, and oftentimes not fit for purpose.

The main lens into this discussion is the weather itself. The weather is constructed as catalyst and force in shaping the ways its study evolved in colonial Malaya, with its supposed stability, for example, precluding any major investment in forecasting or research into atmospheric phenomena as was the case in monsoonal India or typhoon-prone Hong Kong. The resulting science emerged in response to the interests of enthusiasts and on an "as needs" basis, rather than as an essential component of imperial strength. In this way, these chapters set the scene for the book's main premise, calling for an approach to the history of meteorology that asks for a greater appreciation of nuance and of the role of specific climates and environments in reading the scientific past.

STRATEGY AND EXPERIMENTATION IN THE STRAITS SETTLEMENTS, C. 1800–1900

In 1786, Francis Light negotiated the occupation of Penang Island on behalf of the British Crown, marking the first formal incursion of British power into the Malay Archipelago in the pursuit of a strategic positioning to rival Dutch power in the region. Light had stepped into a complex political situation, and the takeover saw the culmination of several years of political maneuvering. First, there was the desire of EEIC officials to establish a foothold in the region after failed attempts in Aceh and Riau, on the north Sumatran peninsula. Second, Penang Island fell under the jurisdiction of the Sultanate of Kedah, on the Malay peninsula. Successive sultans of Kedah had been keen to offer the British areas of the coast in return for a military ally in their power struggles with Bugis rivals on the peninsula and later, as incentive, to offer protection as the prospect of war between Siam and Burma became imminent.[10]

The British saw strategic and political potential in the island that stretched far beyond the immediate balance of power in the region. The internal politicking was merely an inroad. "The climate of this Island is mild and salubrious to a degree which, when its near vicinity to the equator is considered, appears astonishing," remarked Norman Macalister in 1803, in his justification to the EEIC of the importance of Penang to British interests. "It is never liable to extremes of heat and cold," he continued, and "seldom to violent, and never to continued, rains, such as are common on the coasts of India. . . . On the whole it is impossible to conceive a finer climate than that of Prince of Wales Island."[11] In 1822, Nathaniel Wallich, then superintendent of the Botanic Gardens at Calcutta, visited Penang Island, Malacca, and Singapore. Of Penang, he wrote to Charles Lushington, secretary to the government at Bengal, how the climate would be very conducive to the growing of economic crops.[12] Such descriptions of the climate touched upon vital areas of colonial interest: the expansion of trade through the exploitation of natural resources and the ability for Europeans to settle comfortably and healthily.

There was little in the way of science on which to base these observations; they were predominantly experiential comments. Tropical meteorology—defined here as a field with a focus on cyclones, the monsoon, and teleconnected circulation patterns—was nonexistent for the peninsula.[13] What research had been undertaken was predominantly because of early explorative or military expeditions or in connection to disastrous events in tropical regions. After the establishment of the Royal Society in 1660, for instance, a major disaster off the coast of Guadeloupe in 1666 resulted in the destruction of an English invasion fleet. The disaster brought attention firmly toward seeking a better understanding of the nature of winds and tropical storms, largely due to pressure from government and merchants for whom such losses had major financial implications.[14]

Likewise, ventures into Asian waters stimulated interest in understanding the laws governing storms. Robert Boyle's and Robert Hooke's research on barometers' responsiveness to atmospheric pressure enabled scientists to study storms in more detail and there was a new array of information on winds, tides, and rains generated from trading voyages across the Indian Ocean World to the East Indies and to China, such as Captain Joseph Huddart's *Oriental Navigator*.[15] The study of the trade winds, the monsoon, and the extremes of tropical cyclones were to dominate research in the region, albeit with significant disagreement among people such as Edmond Halley through to Reid and Redfield, or later, Henry Blanford, Charles Meldrum, or Ralph Abercromby.[16] As James Horsburgh noted in 1843, "to be able to prognosticate the coming of these tempests would be very useful to navigators, but this cannot be done with certainty, for they frequently commence without giving much indication of their approach."[17]

The better understanding of these phenomena was an integral reason for British sponsorship of scientific observatories in Asia, including the Hong Kong Observatory in 1883, just as major storm events off the coasts of French and British territories acted as catalysts for investing in meteorological services in those countries.[18] The Black Sea Storm of 1854 and the Royal Charter Storm of 1859 are two obvious examples, both forcing a shift in policy and funding for the nascent storm warning services in both countries. In France, the scientific policy of the Paris Observatory changed, kickstarting a process that would eventually underpin their meteorological network for the protection of regional maritime interests.[19] In Britain, the 1859 disaster gave impetus to Admiral Robert Fitzroy's campaign toward a meteorological office as a formal government-funded national institution.[20]

The Malayan peninsula, however, was a region unaffected by cyclones and with less pronounced monsoonal impact, and few resources were spent on undertaking meteorological research for its own sake. There was simply little incentive. Early observations were usually undertaken as a by-product of or in conjunction with other investigative surveys. Francis Light's takeover of Penang Island, for example, resulted in some wind and general weather observations being made at Fort William, the EEIC's newly established military base, in 1789. These are probably some of the first terrestrial instrumental weather observations to have been recorded for the peninsula, though many British ships had called at the island for short periods before, leaving descriptive accounts of the local weather and climate.[21] As the British consolidated their stranglehold across the peninsula, similar sets of records were obtained in Malacca in 1809, 1815, and 1816; Penang Island in 1815–1816 and during the 1820s; and in Singapore from 1820 to 1824. These observations were all made at central and visible sites in the early urban settlements, originally at military forts but shifting quickly to being made at civic or governmental buildings, including Government House (Penang and Singapore) and the library (Penang). Gradually the number of meteorological registering stations increased as the towns expanded, although the stations remained predominantly urban. These early investigations of tropical climate centered on establishing general patterns and trends in the climate and considering how these might impact on agriculture or on health, if at all.

Two competing narratives of Malaya's climate emerged during this period and coexisted. One emphasized the peninsula's unchanging atmosphere and the other concerned an emerging discussion on precipitation variances at seasonal, annual, and multiyear scales. The dichotomy was not fully black-and-white, but there was a tendency for the former opinion to persist across popular and untrained circles and the latter opinion among people who, albeit not scientists, spent time in positions where the weather was important or where there was access to statistical data. Thus, for example, it was common to hear views expressed such as those of Charles Burton Buckley—newspaper editor, writer, and civil servant—who described the commonly accepted theory that "the rain in Singapore falling in showers throughout the year, and not confined to one season, gives a perpetual verdure to vegetation." The local press advertised how "those alternate changes of temperature and seasons from heat to cold, and from months of dryness to months of constant rains . . . are unknown in the 'Straits.' Here there are no seasons."[22]

On the other side of the fence, a more nuanced picture was painted by Captain James Low of the Madras Army. Low had spent five years in India and extensive periods in Thailand prior to his arrival in Penang. His time in Southeast Asia resulted in the creation of land surveys, maps, and ethnographic, historical, and language studies of the region produced during his several different military and civil postings.[23] He produced a detailed description of the local environment and climate in which he acknowledged periodic droughts, seasonal rainy periods, and small differences among the climates of Penang, Province Wellesley, and Singapore.[24] Such differences were also pointed out by Dr. T. M. Ward, also of the Madras Army, whose medical training and experience with medical meteorology led him to conclude that "Province Wellesley is a healthy station, and that its climate seems to correspond with that of Malacca. It does seem on the whole, certainly, more healthy than the Island of Penang."[25] Writing in the 1830s, both men made use of the small amounts of meteorological information that had been collected thus far in each settlement, rather than necessarily undertaking new observations.

One of the best descriptions of the climate during this early period was furnished by James Richardson Logan, described as an intellectual of a retiring nature who devoted his time—outside of practicing law in Penang—to study and research. In 1847, he founded a journal that was a natural extension of his personal interest in the geology, geography, and philology of the region, the *Journal of the Indian Archipelago and Eastern Asia*. Logan contributed many articles of his own, as well as adding commentary and responses to articles published by his peers. The journal became a huge hit among local audiences with writers contributing articles from across colonial Asia and enabling regional research and views to be circulated at an apparent "time of great scientific power and literary activity in the Straits."[26]

While not making his own weather readings per se, Logan utilized other available information with great alacrity. Bringing together the information gathered thus far by military and medical officers, including Horsburgh, Low, and Ward, he carefully calculated statistical means of temperature and pressure and numbers of rainy days, highlighting seasonal differences and extremes. Droughts, he observed, occurred every five or six years in Penang, whereas floods were more frequent and sporadic. Like Low and Ward, Logan pointed to differences in climate across the settlements, which he derived from his analysis of temperatures, wind, and humidity and, significantly, from a very detailed reckoning of the readings made in Singapore between 1820 and 1824 by the EEIC Bengal Native Infantry officer Charles Davies. This data revealed a more variable and less predictable climate than had perhaps been allowed previously. Based on these studies, Logan argued the case for longer-term continuous weather observations, "before we can safely reduce our experience into general laws . . . [as] the irregularities . . . are so great."[27]

It is perhaps something of a shame—as Logan himself noted—that he did not have access at that time to the readings made at the Singapore Observatory between 1841 and 1845.[28] This four-year set of continuous observations would in many respects have fulfilled his call for multiyear analyses, if only for the one location. The Singapore Observatory was a short-lived experiment in investigating geomagnetism as a node in the global "magnetic crusade" of the 1840s.[29] Headed locally by EEIC lieutenant Charles Elliot, the initiative to establish a temporary observatory building on the intersection of the Kallang Road and Kallang River was supported by the combined forces of the BAAS and the Royal Society and was enabled through the logistical networks of the EEIC. Aligned with the cause of establishing laws and theory "based on known facts," the geomagnetic observatory network worked on the nineteenth-century principle of data gathering and statistical analysis to understand natural phenomena.[30] The project was headed officially by Edward Sabine at the magnetic department at Woolwich, London, and by Humphrey Lloyd at the Dublin Observatory. Observatories were set up in Toronto, St. Helena, Van Dieman's Land, Madras, the Cape of Good Hope, and Singapore. They were staffed by military officers under the Royal Artillery, Royal Navy, and the EEIC, with local assistants hired to make twelve readings per day at two-hour intervals for six days a week. The project was ostensibly an investigation of magnetism, but observations of temperature and pressure were taken as a matter of course. Magnetic measurements were affected by alterations in the atmosphere, and careful monitoring of the external conditions was considered critical to accurate magnetic readings.[31]

The Singapore Observatory was small but well appointed, despite a lack of funding. For weather readings, there was a dry and wet bulb thermometer and a barometer. The fifty-six-by-twenty-eight-foot building that housed the observational equipment was well ventilated, surrounded by an open veranda, and care was taken to not contaminate the local conditions by, for instance, burning nightlights or even incorporating an observer's body heat. Elliot lived in an adjacent house, so that he could always be on hand, and he worked with European assistants and a "native" writer to ensure that the frequent readings continued to be taken at all hours.[32] The venture was riven with problems, however. Elliot was at the mercy of funding restrictions that disqualified him from expanding his observational work across the peninsula and into East Malaysia, which, he argued, would have greatly benefited the study of tropical conditions.[33] Although he managed to spend two months making observations in Borneo, this expedition was largely at his own expense and was severely hindered by a lack of staff to carry out operations.[34] The eventual full withdrawal of sponsorship forced the operation to close entirely, a decision that Elliot contested vehemently.[35] Toward the end of the period of the observatory's operations, he also suffered a blow in that "a robbery was perpetrated . . . and the thief succeeded in carrying off the dipping needle—a loss which, as it cannot be supplied for many months, will interrupt the admirable series of observations in progress."[36] The "dipping needle" was a magnetized needle, part of the inclinometer used for measuring the geomagnetic field, and was thus critical to the observatory's study of magnetism.

Although Elliot's observatory was lauded by some contemporaries as a great and worthwhile operation, not everyone felt the same way. Some doubts were cast over the accuracy of his observations.[37] Showing the racially motivated skepticism concerning non-European workers of the age, the Madras astronomer William Jacob—who was responsible for the observations' publication—in a letter to the Cape astronomer Thomas Maclear in 1850 stated, "We are just now printing the Singapore observations, the whole series of 5 years; they are not first rate, I think, much seems to have been left to the Sapper assistants, who doubtless loved a drop when they could get it and neglected or fudged their work."[38] Dr. Robert Little was working in Singapore at around the same time as Elliot and also had things to say about the work undertaken at the observatory. Little was interested in the relationship between weather and disease, and he had constructed a medical topography for that town, some of which was based on Elliot's meteorological observations. His account—often tending toward the descriptive—concerned ideas on how climate affected the body, and he incorporated popular as well as scholarly thought. Considering the *angin Java* (Java winds, a type of squall common during the southwest monsoon), for example, he noted how the "native" and European populations alike were fearful of these "most unhealthy" winds, an idea that he refuted as superstitious nonsense.[39] In 1849, Little and Elliot were to engage in a heated public debate over the weather observations made at the observatory, one into which James Logan also waded.

Little's medical topography utilized the observations of rain and temperature made at the Singapore Observatory from 1841 to 1845. Little was dismissive of Elliot's methods, suggesting that the instruments had not been kept in such a way as to avoid temperature contamination and pulled his concerns into a missive that was published in Logan's journal. Its publication provoked an angry response from Elliot, which Logan also printed, the same year as his own assault on Little's work. Elliot first stated that Little had deliberately set out to antagonize him, Little having also apparently forwarded his critical comments to several scientific societies. Elliot was unsurprisingly concerned therefore that "some of the members may imbibe totally erroneous opinions regarding my observatory and the trustworthiness of the observations."[40]

Elliot's rebuttal, while clearly motivated by hurt pride, does problematize Little's scientific voracity. Stating that Little's own weather observations "existed nowhere but in the writer's imagination," Elliot went on to point out that it was impossible that the doctor could have made an accurate assessment of his methods and instruments before critiquing them as Little had only visited the observatory "but once, and then after his paper had been written." Further, Elliot stated that Little's own scientific research was erroneous, born out of an ignorance of basic meteorology and over reliance on Aristotelian (and thus outdated) law.[41] Elliot then also attacked Logan as editor for supposedly defending Little's critique of his work (Logan had made some comments suggesting that Elliot's observatory was not well run), angrily announcing that Logan just agreed with "anything the Doctor has asserted, and with as little reason; if the Editor, instead of laying down the law in Commercial Square, would have just taken the trouble to visit the Observatory he would have seen and judged for himself how unfounded were the assertions which had gone before." He continued in this scathing attack on Logan's scientific credibility, suggesting that Logan's temperature observations on which he based his 1848 study were not credible. If Logan had at least come out of his journal offices, Elliot fumed, he would have realized that his observations were entirely impossible: "where did the editor ever find 95° in the shade?" If such temperatures were accurate, Elliot argued, "one half [of the population] would be driven from Singapore by fits of apoplexy and the other half would be protracting a miserable existence by sitting in their offices with damp towels round their heads."[42]

Elliot's comments about Logan's imprudent support for Little may have been exaggerated, because Logan had indeed also published a harsh critique of Little's medical topography that same year. It is patently apparent that there existed many differences of opinion as to how meteorology was and should be practiced.[43] The episode is revealing, then, not simply of local rivalries (though that story is interesting in itself) but also of the state of contemporary meteorological knowledge. Basic standards of observation were unsettled, and there were few established frameworks for organizational or institutional practices either globally or locally. Neither was it considered possible to create an ideal environment for science outside of European climates. As Logan himself stated, "a Thermometrical Observatory which would be free from defects in England, may not be adapted for Singapore, where the heat during the day is comparatively so great."[44] He thus perhaps did not hold Elliot fully responsible, and true to comments regarding Logan's nature that he was both a charming and a retiring man, he attempted to smooth the situation by suggesting that the circumstances were simply against the possibility of constructing the "perfect" observatory.[45]

The episode also illustrates some gaping holes in scientific method. From its establishment in 1831, the BAAS had called for the global collation of weather observations so that patterns could be deducted and general laws for climatic phenomena established. In a general report on the state of meteorology in 1832, the physicist and glaciologist James Forbes argued the vital importance of going beyond the current "irregular and far-spread surface . . . spots chosen without discrimination and treated on no common principle" when it came to collecting readings, as well as avoiding "the construction of some arbitrary general laws to connect phenomena." But further, he contended, trained scientists should be released from their "constant occupation" and "mechanical task" of observing, so as "to be able to devote their time to pursue inquiries of interest both to meteorology and its connection to other sciences."[46] Thus a distinction was made between the scientist and the observer with the level of skill that was needed for each role. Forbes's claims were couched in the understanding that, although some progress had been made in British and European meteorological research, such progress only exposed what was left to be done.[47]

This was even more of a problem in British colonies where, with the possible exception of India, there were not "a sufficient number of trustworthy observations, in order to obtain results of primary importance to the science, and which may one day enable us to arrive at the true form of the daily and annual curves of mean temperature with a precision almost mathematical."[48] Forbes had great hopes that the observations made via the magnetic experiment, including those made in Southeast Asia, would go some way to filling the gaps he highlighted, though he noted that "no material step whatever has been made" on the subject of climatology nor on the seasonal variation of global isothermal lines.[49] It is a great shame that Elliot's observatory closed after only four years.

GLOBAL FERVOR, LOCAL ENNUI

The 1850s and 1860s were a vibrant era for meteorology globally. Across much of Europe, America, and Prussia, the fields of storm tracking and forecasting were bolstered by the new telegraphic technology, which enabled observations to be transmitted at speeds and distances never known before. For example, Matthew Fontaine Maury's work toward advancing a dynamical meteorology—in promoting a nationwide American observational network and in helping to promote the first International Maritime Conference held at Brussels in 1853—set a precedent showing the possibilities unlocked by international collaboration and how in practical terms networks could be established at wider scales. Alexander von Humboldt's and Heinrich Dove's separate work on isotherms led to the application of statistical method in establishing global, latitudinal norms. Dove championed work investigating anomalies, such as nonperiodic long-term temperature changes, just as Adolphe Quetelet was doing for nonperiodic barometric variations. These he visualized graphically as isolines, thus pioneering isobarometric geometry by identifying isochronic lines of atmospheric pressure.

Much of this work could have been turned to debunking the idea of climate stasis in the Straits Settlements, yet this was not to be.[50] Although colonial officers appeared to be aware of many of the scientific developments overseas, this new knowledge was not always applied in situ, perhaps for two obvious reasons.[51] First, the men involved in meteorology in the Straits Settlements were not scientists but colonial officials, military officers, and gentlemen observers, Renaissance men with their fingers in many pies. Second, there was a lack of instrumental weather observations available upon which to work. Most observations on the peninsula and its islands were made in a piecemeal fashion, despite efforts by the Meteorological Committee of the Royal Society for all colonies to furnish the Colonial Office with weather diaries twice a year.[52] The norm for the Straits Settlements, or indeed Britain's Southeast Asian forays more broadly, had been short records such as those made by Henry Burney while stationed at Tavoy, Burma (now Dawei, Myanmar) during November and December 1827 or the occasional scattered observations that were found for Singapore for a few months in 1833, 1835, 1837, and 1840.[53] Elliot's four-year record from the 1840s Singapore Observatory was something of an anomaly by the standards of the time and place.

In the 1850s and 1860s, the situation declined further. After the Singapore Observatory closed in 1845, meteorological study appeared to have descended into a dark age, with little evidence of any research or record-keeping in the towns until the late 1860s, and this was both piecemeal and only for Singapore. The only exceptions are a short dataset for Alor Gajah near Malacca made during the 1850s and some readings made at Singapore's Horsburgh and Raffles lighthouses during the 1860s.[54] Heinrich Dove would not have been surprised. Across colonial Southeast Asia, only a few places including Batavia and Samarang (Semerang) issued any regional observational data to overseas societies during this time.[55] This may not have been the actual sum of observations (there were weather journals kept by plantation owners, for instance), but it is certain that few official weather records were kept at government or scientific institutions or were printed in newspapers.

The plantation owners are something of a special case, and surviving records seem to exist only for Singapore. These records were made by men such as George Henry Brown or Jonas Daniel Vaughan, plantation owners at the Mount Pleasant and the Killeney Estate respectively.[56] That these individuals had weather interests comes as no real surprise, but the (mostly) abundant rainfall meant that a formal investment in meteorology even for important agricultural pursuits likely did not hold much persuasive power at governmental levels. For much of the century, agricultural research was in the hands of the Agricultural and Horticultural Society, later the Agri-

Horticultural Society, a nongovernment group of social elites including plantation owners who funded their own information gathering.[57] Indeed, when in 1886 the Resident Surgeon called for all Collectors to combine agricultural information (such as when rice *padi* was sown and harvested) with meteorological returns, the request fell on deaf ears.[58] Where money was invested in agricultural research, this was predominantly into pests and crop diseases. The ruin of two potentially major commercial crops—nutmeg and coffee—during the nineteenth century was testament to where scientific research monies were considered needed.[59] The lack of weather reporting in Malaya is also supported by inhabitants' comments in the local press. Writing under the pseudonym "Delta" in 1877, the local businessman and political commentator William Read lamented the absence of meteorological research in the colony over the previous decades, attributing the situation to yet "another symptom of the spread of Dundrearyism, and the reign of somnoleus Dullness over all the Departments of Government."[60]

Things were changing, but very slowly. During the 1850s, it became apparent to many colonial officials that deforestation for plantations needed to be far better controlled and more discriminating of the type of forest being cleared.[61] New legislation, often following the example set in India and Burma, was deployed to avoid the nascent problem of soil erosion on steep hill slopes, but there was a conterminous narrative that rampant deforestation might cause local changes in rainfall.[62] The idea of human-induced localized weather change had developed out of a combination of local circumstances and international knowledge, with deforestation under investigation as a possible cause of declining precipitation. Work on desiccation was pioneered in Europe and Prussia by scholars including Humboldt, Quetelet, Peter Kropotkin, and Dove, using a combination of correlative observation and new research into atmospheric changeability. To understand whether rainfall was changing in the Straits Settlements, longer-term continuous datasets of observations were critical, a fact that "Delta" made clear in 1877.

In 1869, the Straits Settlements' administration officials made a first move toward keeping better records by employing hospitals as registering stations, equipping them with basic measuring instruments including thermometers and rain gauges, and training medical assistants to take readings. This effort came under the operational jurisdiction of the Medical Department and the colonial surgeon or the principal chief medical officer (PCMO) who became the formal head of meteorology for the colony. Singapore's Kandang Kerbau Hospital became the coordinating point for this venture, and it was fitted with a small meteorological observatory. The new direction was enabled by a critical change in the administration of the colony: the transfer of powers over the Straits Settlements from the India Office to

London in 1867. The change separated many state and commercial concerns and awarded more autonomy to local colonial governments for their spending and decision making. As Lenore Manderson has argued for the medical service, the transfer enabled local authorities to invest more time and money into civil institutions.[63]

Placing meteorology into medicine was not unique to the Straits Settlements but was common practice across the British Empire and in Europe where "medical meteorology" was a recognized field of study. It was considered especially important for study in the tropics, where extremes of rain, heat, and humidity were thought to push the (European) body to the limits of its endurance. As David Arnold has shown for British India, the colonial gaze—as exemplified in the medical discourse of men such as James Johnson, James Ranald Martin, and Robert Scoresby Jackson imagined disease as being somehow more "acute" in tropical locales.[64] The new pathological and atmospheric climates Europeans were now inhabiting in Asia were thought to exacerbate or cause disease, and money was pushed into the medical study of the weather, even where it was not studied for the sake of meteorological science. Norman Lockyer and William Hunter (the Director General of Statistics to the Government of India), for example, investigated sunspots and their relation to cycles of the Indian monsoon, famine, and cholera.[65] Doctors and men of science based in India were also investigating through statistical correlation of weather records with disease outbreaks whether epidemic years (especially epidemic cholera) were characterized by any peculiar meteorological phenomena.[66] Henry Blanford, the first director of the Indian Meteorological Department (IMD) was able to critically address Lockyer's ideas, for example, by using newly gathered observations for the El Niño–inspired droughts of 1877 and 1878 that devastated India, China, Australia, and parts of Southeast Asia and comparing the drought incidence with outbreaks of cholera.[67]

In the Straits Settlements, cholera was also a major concern of the medical establishment, who saw droughts as precursors of the disease.[68] To have on hand multiyear daily weather records, rather than rely on memory or patchy annual means, meant that cyclical patterns of weather and disease could be better identified, and this, eventually, as theories on disease and disease vectors evolved, led to more nuanced interpretations of the intersection of weather and ill-health. Following Blanford's example, Dr. F. K. Hampshire, colonial surgeon at Penang, also studied the 1877 drought and its impacts on that settlement, arguing that it was not the drought per se that caused the steep rise in cholera that year but the "deficiency of wholesome drinking water." He also brushed aside arguments suggesting that the upsurge in smallpox that year had anything to do with the weather.[69]

THE UNCHANGING CLIMATE

Notwithstanding the institutionalization of meteorology in the Medical Department, the science still suffered from a lack of a dedicated central body, despite promising examples from elsewhere in Asia, either British or otherwise. Observations had been made at an observatory in Mauritius since 1831, the IMD was established under Blanford in 1875, and the Hong Kong observatory under the directorship of William Doberck in 1883. It was the Jesuits, however, who led the way in cutting-edge research for the region. The Peking Observatory in China had Jesuit directors and had counted meteorological studies among their astronomical work since the 1700s. From 1873, Jesuit fathers had made meteorological observations is Shanghai. In the Philippines in 1865, they established one of the first observatories in Southeast Asia, led by Fr. Federico Faura, who with his successors conducted important research into typhoon warning systems and instruments and by the end of the nineteenth century led the way in regional typhoon studies. In Vietnam, the French Jesuits also built an observatory that opened in 1906 at Phu-Lien, near Hai-Phong, although they had begun making observations two years earlier under the watchful eye of its first director, Dr. Le Cadet, a renowned scholar who coordinated research into atmospheric phenomena, geomagnetism, and other studies.[70] In Japan, the Tokyo Meteorological Observatory was established in 1875, followed by another site at Osaka in 1883.[71] The Dutch founded an observatory at Batavia under the directorship of Johan Maurits Mohr and had been pioneering local scientific discussions under the Bataviaasch Genootschap van Kunsten en Wetenschappen (Batavian Society of Arts and Sciences) since the eighteenth century.[72] By the nineteenth century the Dutch East Indies colonists were leading the way in many scientific areas, including tropical botany and agriculture, using the mathematical and empirical tools of contemporary science.[73]

Therefore, while malaise was certainly not limited to Malaya, the Malayan situation seemed to fare worse than elsewhere, British-led or otherwise, by the late nineteenth century. Subsequent to the closure of Elliot's observatory, no new observatories were founded, so for much of the mid to late nineteenth century, Malayans had to content themselves with the small Kandang Kerbau Hospital facility, a second small site at an early signal station and military base at Mount Faber in Singapore, and some registering stations in urban Penang and Malacca.[74] Henry Ellis, master attendant of the Navy based in Singapore, repeatedly called for the necessity of a proper observatory across the 1870s and 1880s, especially after he saw the British government pay for the new observatory in Hong Kong in 1883.[75] Another small facility was eventually set up on the harborfront in 1893 under the jurisdiction of the Harbour Master's Office, but there was nothing comparable to the institutions in Hong Kong or in India.[76]

Harking back to some of the depictions of the tropical climate by contemporaries, the sense of a stable climate in the Straits Settlements, albeit one that would gradually dissolve as time went by, perhaps removed any sense of urgency or necessity. The observatories in Hong Kong and India were built with the idea of better forecasting typhoons, providing a storm warning service, and studying the monsoon. There were no typhoons in Malaya, and the effects of the monsoon were considerably less. It was only after longer-term observations became available in 1869 that patterns for flood or drought periods could start to be identified in any scientific way, and this was a science of patience, with a long wait for enough years of data for meaningful information to emerge. As Helen Tilley has argued more broadly about imperial scientific institutions, they "occupied an interstitial space that was neither national nor international," operating on an "as-needs" basis within locally specific contexts.[77] The lack of formal governmental support and funding for weather science could well have stemmed from the lack of major nature-induced disasters, therefore.

In the early 1800s, weather research projects were ad hoc and funded by third parties (the BAAS and the Royal Society, for example), but sizable coordinated projects on national scales relied on more reliable, continuous sources of funding. Even after power was transferred from the EEIC to the Colonial Office in 1867, meteorology was not funded for its own sake. Rather, it was embedded into medical research and structures and was restricted to recordkeeping and correlative analysis. Outside of the medical establishment, meteorology was studied only at port authorities and by plantation owners, and in each case, little more was expected than basic recordkeeping and forecasting.

Thus, although more registering stations were established over the last two decades of the nineteenth century, research into climate and weather did not extend much beyond correlative and statistical analysis. This neglect had a clear influence on the local status of the science, as the authority for weather reading was moved away from the medical service and into the hands of the ostensibly unconnected Museums Department in the late 1910s.[78] This could have sounded the death knell for Malayan meteorology were it not for new champions and a shift in governmental priorities.

CHAPTER 2

MAKING METEOROLOGY MAINSTREAM

WEATHER SCIENCE IN EARLY TWENTIETH-CENTURY MALAYA

With the twentieth century came a shift in the way that meteorology was viewed and practiced in Malaya, in sharp contrast to the nineteenth-century situation described in the previous chapter. The reasons for the change in status were embedded in new local and global realities and the advent and availability of technologies and methods that connected to, or necessitated, forecasting. Meteorological forecasting was considered important in the nineteenth century, but the rise of commercial and military aviation placed new demands on meteorological services and required a redirection of national government's scientific priorities. The new wireless, radar, and radiosonde technologies of the 1900s enabled meteorology to fulfil these demands.[1]

In Malaya, transformations in scientific practice were also transformations in spatial practice. Meteorological spaces—such as observational registering stations—moved out of medical facilities and became independently controlled by new dedicated meteorological institutions staffed by trained meteorologists. The main stations and observatories remained predominantly urban, mainly based in Singapore, but the network of registering stations expanded geographically across harder-to-reach rural, jungle, and mountainous areas, a shift that was made possible by advances in communications technologies and infrastructures. Forecasting over the kinds of distances required by international flights also necessitated improved networks and communications not simply across the country but across regions and services in military, scientific, and civil realms, a subject that dominated conversation at the new raft of international and regional meteorological conferences now taking place in Asia and Southeast Asia. Meteorology was likewise reaching into the sky, as aviation and the radiosonde propelled upper atmosphere research to new heights.[2] Meteorological science evolved at scales geographic, technical, and professional. As with the previous chapter however, such developments ought not be considered inevitable or systematic, the science remained subject to restrictions in funding, resources, and need.

This chapter therefore considers the Malayan experience as part of a broader British imperial meteorological agenda but one that was peppered with local specificities. Beginning with a discussion of how an international war and a nascent aviation industry catalyzed the formation of a dedicated meteorological service in 1929, the chapter will then explore how that newly formed service interacted within a global arena and knowledge space across the 1930s until the outbreak of war in 1939, concluding with a look at how the Japanese occupation of Malaya from 1942 to 1945 impacted that service work and research. Throughout, close attention will be paid to the multiple factors involved in expanding, or hindering, meteorological science research in Malaya.

EARLY TWENTIETH-CENTURY METEOROLOGY, WAR, AND AVIATION

According to Robert Friedman and other scholars of meteorological history, the First World War was a catalyzing moment for weather science in Britain.[3] In 1914, German government officials were in possession of an advanced and organized weather service, far superior to that of the British. This was in large part because of their early adoption of aeronautic technologies, with the German town of Leipzig already a leading center of aircraft manufacturing by the 1910s. Improving weather knowledge and forecasts were essential to flying the early rigid Zeppelin and light aircraft that operated at the mercy of the atmosphere, especially the squall lines—linear thunderstorms that develop quickly at the edge, or ahead of, a cold front.[4] It was a squall line that caused Germany's first major loss of life in a Zeppelin crash in 1913, for instance.[5] Air raids, nighttime flying, and even some field operations—including barrage gunnery, gas warfare, and long-range artillery bombardments—demanded reliable, real-time forecasts from manned in situ observational balloons and airplanes that relayed the most up-to-date information possible to military operations' planners. Germany's advanced weather service left the British forces at an initial disadvantage, one that became increasingly apparent as the war evolved. In 1915 the British responded by establishing a new field-based Royal Engineers unit to focus on weather services for the war effort under the supervision of the British Meteorological Office and its then director, Sir Napier Shaw.[6]

Although this war was not fought on the ground in Malaya, there were ramifications for meteorology in the colony as aviation and military concerns developed.[7] In 1917, Napier Shaw was asked to report on Britain's overseas weather services for the Ministry for Reconstruction. He was quick to point out how the Army and Navy had little specialized data for regions beyond western Europe, so it was often a case of improvisation when needs arose. For the most part, the collection of weather information and—importantly—the sending of this information to Britain from the colonies was an ad hoc affair. In the Straits Settlements, for example, the Colonial Secretary asked that the British Resident at Kuala Lumpur arrange to send the Medical Department's full annual meteorological report to the Meteorological Office and the Royal Meteorological Society in the 1890s. This was in addition to the abridged information that the Resident's office already sent annually in the form of monthly or annual means. But as all these returns were sent to Britain several months after the end of a year, they served no useful purpose in terms of forecasting.[8] Even by the outbreak of the First World War in 1914, there was no standardized real-time reporting system in place within or from the Far Eastern colonies: everything was retrospective. This circumstance yet again put Britain behind Germany, which had, in Shaw's opinion, a better coordinated body of global weather data and superior meteorological training programs at their universities.[9] This led the British authorities to try and improve the flow of information across its imperial world, something that was to become significantly more important as Singapore became a focus of British military strategy in the Far East, following the lessons of the Great War.

During the interwar years, the British realized the necessity of changing strategy as the balance of power in the region changed. Pivoting away from Japan, the end of the Anglo-Japanese naval alliance and the signing of the Washington disarmament treaty in 1922 were designed to protect American and British interests in the event of armed conflict between the United States and Japan, just as Japan was emerging as the most likely threat to Britain's naval superiority in the region. This necessitated a permanent British naval presence in the region hinged on Suez, Hong Kong, and Singapore.[10] Officials at the British Admiralty argued vociferously for a new naval base to be built at Singapore, on grounds of the defense of strategic outposts and supply lines from Asia to the Middle East, and it was completed at a massive cost in 1938.[11] But the future of aircraft was connected to the same argument as for the future of Far Eastern naval capacities during the interwar period, especially in view of the new specter of regional aerial attacks. Indeed, the rise of air power proved a useful means to argue against further increasing expenditure for naval capacities.[12]

As the First World War ended, a vision of a future world linked by an air transport network was at the point of consciousness.[13] The legal framework for the internationalization of air traffic based on a (Western) concept of national airspace was set out in the Convention relating to the Regulation of Aerial Navigation (1919), ratified during the postwar Paris Peace Conference. This convention also recommended guidelines for the international exchange of weather data, which was augmented by the newly minted International Commission for Air Navigation, intended to oversee meteorological standards.[14] These new initiatives buttressed the British Imperial Airship Scheme, the brainchild of aeronautical engineer and politician Charles Dennistoun Burney. His 1922 proposal (known as the Burney Scheme) called for six large commercial airships to service the empire, flying from Canada to New Zealand and back again. Many politicians and business representatives recognized the commercial potential of the proposal including the managers at Shell Oil, who saw this as an opportunity to dominate the global airspace market.[15] This was the future of transport. In 1924, Britain's Labour government authorized an adaption to Burney's plans so that they, not Burney, would have control. The government approved building two airships and renamed the plan the British Imperial Airship Scheme. Technicians at Burney's Vickers Company were to design the two ships, the R100 and the R101, which would be constructed at the government-owned Royal Airship Works at Cardington. The future of airships would soon be curtailed, but the door to the possibilities of faster transoceanic travel had been opened.

As the British Imperial Airship Scheme was rolled out, the excitement in Malaya was palpable.[16] Aviation was incredibly appealing to expatriates who envisioned a future with shorter and more regular journeys between Southeast Asia and Britain, not to mention the convenience of faster postal services and the ability to import a wider range of fresh foods from across the region. In 1926, Mr. Alan Cobham of the Singapore Ex-Services Flying Association declared at their annual dinner how "as in the past Great Britain had been a great nation of seamen, so in future she must become a great nation of airmen . . . already there are unmistakeable signs that at no distant date flying will be brought within the reach of everyone of moderate means."[17] His sentiment was echoed in the Malayan presses, which published reprints of international articles alongside locally produced pieces heaping praise onto the airship scheme and other forms of air travel. Extolling airships' technological innovativeness and luxury, across 1927 and 1928, for example, the papers described the "wonders" of the four-story high dirigibles, calling them the "future" of transport and debating enthusiastically on how journey time between Britain and India could be cut to a week and to Australia to less than a fortnight.[18]

Although dirigibles never ended up flying to the tropics, their space was quickly filled by the airplane and the airliner.[19] The colony was captivated by the daring international exploits of the Wright Brothers, Louis Bleriot, and others, but Malayans would also witness this new technical future firsthand. When Captain Ross Macpherson Smith arrived by air in 1919, it is widely thought to have been the first time that a piloted aircraft had set down on the island and his visit met with great acclaim.[20] The internationally famous flying team Jessie Miller and William Lancaster likewise stopped in Singapore in 1928 after their plane, the *Red Rose*, encountered difficulties over Indonesia.[21]

One of the most famous flying visitors to Singapore was Amy Johnson. Her successful 1930 flight from England to Australia made a stop at the new Seletar Airbase in Singapore en route between Thailand to Surabaya on 17 May 1930 to refuel and repair. Johnson was regarded as a something of an international star. Her visit elicited far more interest "than has been aroused by long distance flyers who have visited Singapore in the past," or so said the local press. Hollywood's fascination with flying machines also took hold in 1930s Singapore. RKO's new musical featuring Fred Astaire and Ginger Rogers, *Flying down to Rio* (1933), and *Only Angels Have Wings* (1939) starring Cary Grant were played to admiring Singaporean audiences.[22]

It was no coincidence that the aviation fever of the late 1920s and early 1930s encompassed the launch of the first commercial international air service to connect Singapore to rest of the world. This Dutch service linking Jakarta and Singapore had its first test flight just three months before Amy Johnson's arrival. The flight conveyed eight passengers, including the British consul general in Java, VIP guest Sir Josiah Crosby, from Jakarta to the brand new hypermodern British Royal Air Force (RAF) airbase in Singapore, RAF Seletar. Gushing about his experience, Crosby lauded the comfort and swiftness of his experience, which reduced his travel time from the usual thirty-six hours to a mere seven. What really excited the press, however, was the discovery of fresh strawberries and flowers onboard, picked literally that morning from the hill gardens at Bandung just outside Jakarta. The possibility of bolstering Singaporean diets with fresh produce from Indonesia was not lost on contemporaries, given the limited selection of fresh locally grown fruits and vegetables suited to expatriate palates.[23] The Jakarta–Singapore route would be the Straits Settlements' first international integrated passenger, freight, and mail air service, followed by a Singapore to Penang service from 1931.

Within Singapore, the Seletar Airbase services were soon to be extended to a new, more central airfield at Kallang, opening in 1937. By the outbreak of the Second World War, Singapore was strategically placed to be the essential military defensive hub for the Far East, and in terms of commercial aviation, the intervention of the Dutch through the KLM Company quickly expanded its reach to connect Singapore to Amsterdam in 1933. British

Imperial Airways unveiled a London–Singapore–Australia route quickly in 1934, the latter portion being taken over by Qantas Empire Airways one year later.[24] These were immensely significant changes to regional and global connectivity within a very short period. Certainly, when Cecil Clementi, governor and commander in chief of the Straits Settlements, somewhat prophetically announced in 1931 that "Looking into the future, I expect to see Singapore become one of the largest and most important airports in the world," he was not far wrong.[25]

These new routes could not be supported without current weather information and forecasts.[26] Retrospective publication of data in this context was useless. Unlike the lukewarm support that meteorology received during the nineteenth and very early twentieth century, by the late 1920s it was clear that the Malayan provision for forecasting peninsula wide, or for international flights, was deeply unsatisfactory. When the Great War broke out in 1914, the Medical Department (and army Medical Department) were still managing observations as they had been since 1869, with guidance and direction issued from the Indian Meteorological Office at Simla.[27] But the observations made at hospitals or other civic sites were not being gathered with aviation forecasting in mind, they were aimed more at correlative studies and general information for agriculture.

There is also anecdotal evidence to suggest that these observations may have been poorly coordinated and lacked precision. In 1902, for instance, an anonymous contributor was moved to write in to the *Straits Times* to complain about some of the official records then being reported from the main meteorological station at Kandang Kerbau Hospital in Singapore. A friend, this author argued, regularly made his own weather observations at Bukit Timah (around seven kilometers northwest of the hospital) and found far wider daily temperature variations than the official records suggested.[28] This correspondent may have been unaware of the difference that the peri-urban and plantation locations of the two sites may have made to the temperature, but this was not the only complaint. Another anonymous writer (possibly a government official) suggested in 1907 that recordkeeping at the government sites was inconsistent and full of gaps. Where records existed, it was almost impossible for any interested party to access them in order to undertake studies or analysis or even just to check normative conditions. In 1911 too, the editor of the *Agricultural Bulletin* was to lament the sorry state of the colony's meteorological records (or lack thereof), arguing that they could have been invaluable if only they had been properly maintained.[29] Likewise, in 1917, the *Malaya Tribune* reported how discussions of the comparative rainfall at Penang and Singapore—two places frequently compared for their agricultural possibilities—seldom resulted in anything conclusive because of the lack of reliable data.[30]

This failing cannot be placed entirely at the door of Malaya's officials, as it also reflected a wider malaise in adopting new standards and observational conformity. Britain's Napier Shaw, for instance, pushed hard to make imperial meteorological observations and practices more standardized and for observations to be published on a regular basis, but this was not an easy task. His office arranged a weekly weather report, for example, in which was contained the Réseau Mondial, a repository of imperial temperature, rainfall, and pressure observations. Beginning in 1910, this compilation necessitated a uniformity in measuring conventions that was still unattainable among British colonial registering stations. Napier Shaw tried to fix this by adopting recommendations for standards made by Vilhelm Bjerknes and supported by the Gassiot Committee and by advocating to replace currently used British imperial units with metric so as to bring Britain and Europe in alignment. This was unpopular.[31] One result was that the Réseau Mondial content had to be scaled back. Within two years of launching, it went from a publication containing daily observations to containing monthly and annual observations, in part because weather observations tended to be mailed in (and slowly) but also the fact that the adoption of metric units proved hard to enforce.[32] It was the 1930s before most colonial nations were on board with these new systems, and as Aitor Anduaga suggests for the East Asian meteorological services more broadly, the quest for standardization was problematic because it was as much a question of economics as it was of will. Transmitting weather observations far more quickly by telegram became possible in the 1890s, for example, but transmission costs were dominated by word counts.[33] Thus, the weather services were at the mercy of telegraphic companies, their charging systems, and government budgets.

These global obstructions to scientific coordination were reflected in the Malayan experience. In Singapore, classic imperial measurements—including inches for pressure and rainfall—continued to be used well into the 1920s, but on Penang Island, the meteorological station at Fort Cornwallis adopted the metric system in 1917.[34] This was even before Napier Shaw successfully arranged for the colonies to all have their own identical copies of the official *Manual of Meteorology* in 1919, written to bring some consistency to observation making.[35] Because of the lack of trained meteorologists or a meteorological service, the colony also lacked official representation at any of the early important international scientific meetings and meteorological conferences, including those of the International Meteorological Committee (IMO), which came into being in 1873, or even the Conference of Meteorologists of British Dominions (1919), where the Straits Settlements were subsumed under the leadership of the director-general of Indian observatories. This meant that there were no locally based scientists to act as advocates for standardization or, indeed, for tropical research.

AGROMETEOROLOGY AND THE INTERWAR CLIMATE

After the First World War, the authority for making observations was transferred from the Medical Department to the Museums Department. This seemingly incongruous move may have been inspired by the role of the museum as a library and repository for the natural sciences of the peninsula. It may paradoxically also have been a blessing in disguise. The director of museums between 1908 and 1926 was a man named Herbert Robinson. Something of a Renaissance man, he was an avid ornithologist, explorer, and zoologist. He also had a keen interest in meteorology. He, like many of his peers, was critical of the quality of observations and work undertaken by the Straits Medical Department, suggesting that its staff were ill-trained (they were often the hospital orderly or a dresser who were assigned this task in addition to their other duties), their instruments were not standardized, and their readings were temporally erratic.[36] Robinson believed in the importance of improving the number and coverage of stations across the peninsula, especially in rural and isolated areas, and in the necessity of hiring only specialist, dedicated staff. From 1920, he embarked upon a new scheme to employ navy signalmen as meteorological observers to man the remotest stations across the Malay States, and he personally vetted new areas for development into new registering stations, including a hill station at Gunong Tahan in Pahang, for example, which he visited during a botanizing expedition.[37] By February 1921, Robinson had hired four (European) men who were put on a six-week intensive training course in Kuala Lumpur before starting out for the new Pahang station that April.[38]

Robinson's expansion of the network did not come out of the blue. In the first two decades of the twentieth century, Malaya moved away from traditional export crops such as pepper, gambier, and sugar to become a world leader in pineapple, rubber, and palm oil production.[39] This was intensive plantation monoculture tailored toward large-scale commercial companies and government-run plantations, which were overtaking the frontier spirit of nineteenth-century smallholding. The situation demanded new frameworks and infrastructures and was embedded within what has been called an era of British sociotechnical colonial developmentalism, with roots in Joseph Chamberlain's interventionist approach to colonial development by professionalizing scientific services in the colonies.[40] In Malaya, the bureaucratization of the environment began with the launch of the Forestry Department in 1903 and continued with the Department of Agriculture (DOA) for the Federated Malay States (FMS) in 1905 and the Rubber Research Institute, which opened in 1926.[41]

The same year that the DOA began, Napier Shaw called for the British Meteorological Office to play a bigger role in the collection and analysis of weather data across the empire, because it was essential to supporting imperial commercial agriculture.[42] After the First World War, such calls intensified. Bjerknes noted in 1917 that as "the world is entering a period of food shortage . . . meteorology has a responsibility to support . . . the production of food stuffs by developing and improving weather forecasting."[43] In Malaya, a steep depression in the rubber industry followed the end of the war and then again between 1929 and 1932 as the world hit severe recession.[44] The interwar years were also plagued by the specter of food insecurity, which in Malaya, a colony that imported a large proportion of its foodstuffs, was a serious worry for colonial officials.[45]

In 1921, the new Colonial Research Committee published the first in a series of an important scientific reports for the peninsula.[46] This committee's function was to investigate and promote research that enabled the development of imperial economic resources. The committee's survey of Malaya contained an overview of extant research work, which noted how rubber and tin, rather than food, were dominating research. It was therefore deemed essential that the agricultural sector be assisted by the "systematic investigation" of rainfall.[47] There had been extreme weather episodes in Malaya, as elsewhere, in 1910–1911, 1924–1925, 1931, 1933–1934, and 1935, all connected to some degree with especially strong phases of the El Niño Southern Oscillation (ENSO), which caused flooding or drought. Although Malaya did not suffer as much as other parts of the world during these events, Malayans were impacted by disruptions to food supplies that were caused by regional failed harvests and corresponding price hikes. In 1911, for example, a severe El Niño–related drought sent rice prices skyrocketing long after the drought had ended meteorologically, because of regional shortages. The British colonial government were forced to respond by temporarily shifting their imports from Siam to Burma.

Compounding the problem of flooding and droughts were the wider land-use changes that took place over the last century, transforming jungle into plantation.[48] The indiscriminate land clearance contributed to soil erosion, which, by the 1920s and 1930s, reached levels that were impacting on Malayan agricultural productivity and worsening the effects of heavy rain. In 1938, the dangers of erosion were made abundantly clear during the Conference of Colonial Directors of Agriculture held in London where it was argued that "soil erosion and nutrition are the two most important subjects to be discussed . . . it was only during the past few years that the importance of soil conservation had been fully realized, largely as result of the account of the position in the United States."[49]

Although the statement referenced the Dust Bowl of the American prairies, that experience led many people to fear a similar scenario for some of the colonies, especially those in Africa and in Asia.[50] A. H. W. Weir, who had acted as chief conservator of forests for Nigeria, argued how erosion meant "cash crops" including rubber "were in jeopardy"; Edward Stebbing, who had worked as a forest entomologist for the Indian Forest Service before becoming professor of forestry at Edinburgh University, contended that government must sponsor a study of the "direct or indirect action of erosion on water courses and on atmospheric rainfall."[51] While referring specifically to Africa, their comments were taken seriously in Malaya, which was then undergoing its own erosion crisis. Thus, there was a sequence of compelling and intwined reasons for improving meteorological research: increasing imperial agricultural yields, combating soil erosion, and managing food security, all within an era of increasing bureaucratization of state scientific services that were essential to colonial development.[52]

Things were looking up for weather research. The DOA and Rubber Research Institute both began coordinating their own weather data gathering and conducting investigations into rainfall and soil erosion, but the real breakthrough was the opening of the Malayan Meteorological Service (MMS) in 1929.[53] Plans for the new service were some years in the making. In 1924 Herbert Robinson of the Museums Department, along with George Maxwell, then chief secretary for the FMS, put into place Malaya's first scientifically trained and dedicated meteorological officer, S. G. G. Kelliher.[54] Robinson and Maxwell in close consultation with the Survey Department then came up with a proposal that would allow for the development of seventeen new meteorological stations across the peninsula, all designed to extend observational coverage into rural and mountainous regions and paid for by each of the regional Straits Settlements and FMS governments in proportion to their revenues.[55] Central to this was the development of a new weather reporting system for the Harbour Authorities and a nascent forecasting service for the aviation industry, beginning with the Imperial Air Mail Services into Singapore. Described as "the first serious attempt by any officer in charge of Meteorological work here to place the Meteorological Department on a sound basis and to make it fit to meet modern requirements," the service was intended to reach not only across the entire peninsula but across the whole region by exchanging historic and contemporary information on storms with Asian meteorological services.[56]

Critical to this mission was improving communications infrastructures.[57] The new stations were enabled by the latest wireless receiving equipment, as too the existing stations, including the small observatory on Mount Faber, which was gradually improved over the 1920s under the guidance of A. J. Bamford, superintendent at Colombo Observatory, Ceylon.[58] The stations' primary function was, and always had been, to provide a time service for shipping, but it now housed new astronomical equipment, a traveling micrometer for time determination, and the latest wireless receiving equipment.[59] Although wireless technologies had been available for some time, the improvement of short- and long-range wireless capacity and the establishment of empire-wide infrastructures during the 1920s transformed the technology's potential.[60] The Imperial Wireless Chain, for example, was activated progressively across the 1920s, building in many places on preexisting telegraph cable networks and linking stations from Port Darwin to Singapore through Cairo to England.[61] Despite political contention over exactly how and where the scheme should operate and even who should operate it (the Post Office under direct control of the British government or the private Marconi Company), eight new wireless stations using large-valve transmitters and high-power stations provided for long-wave communications between Britain and the colonies.

The first station opened at Leafield, near Oxford, in 1921, the second at Cairo in 1922, and the final link was made between Australia and Canada in 1928.[62] The newly invigorated Mount Faber Observatory, run by A. J. Wilson, benefited from this service, receiving communications from Colombo, Hong Kong, and Manila by 1928.[63] Unlike the nineteenth century when meteorologists had to rely upon observations that were historic by the time they arrived, wireless communication meant real-time information.[64] This had the potential to enable a Malayan service that could operate as a hub within a regional network, just as the observatories at Hong Kong, Shanghai, Tokyo, and Manila had already achieved several decades previous.[65] Nonetheless, the experience of networked information sharing and knowledge seemed less central to the Malayan services than was the situation in the rest of Asia or, as many historians including David Lambert and Alan Lester have pointed out, the world—although the successive superintendents or directors of the service did participate in the increasing number of global congresses and organizations devoted to the dissemination of scientific knowledge that proliferated during this time.[66]

THE MALAYAN METEOROLOGICAL SERVICE

Not much is known about the staff who ran the new service and undertook this ambitious work over the 1920s and 1930s. We do know that the few senior officers were all British and that some of the assistants and all of the subordinate officers were locally recruited. The man in charge from the very beginning was Superintendent C. D. Stewart, and he worked alongside the meteorological officer S. G. G. Kelliher, until they retired (1938 and 1936 respectively). Their incumbency saw the service grow and split into two departments. The MMS headquarters was based at the Fullerton Building in downtown Singapore; climatological research, enquiries, and forecasting shifted to the Kallang Airbase when that new facility opened in 1937.[67]

One tantalizing photograph exists of the staff who worked at the Fullerton, taken on the occasion of Kelliher's retirement.[68] The photo shows a small team comprising Kelliher and Stewart along with Ho Seng An, Ivory Kedit, C. A. Lea, W. Richards, and Che Ahmad bin Abdullah. Not too much is known about any of them, even the senior officers. Lea stayed with the service throughout the Second World War, eventually becoming director in 1950. He entered into official reports but of his personal life or background there is precious little. Of Che Ahmad bin Abdullah there is next to no information, but Ivory Kedit and his daughter were interviewed by the National Archives of Singapore many years later for their "Communities of Singapore" oral history collection.

Kedit was born in 1901 in Sarawak, the son of an Iban indigenous tribesman. His father worked in Kuala Lumpur as a collector for the Museum's Department, searching the jungles for exotic items and animals that would interest the British. Kedit's own career started when an uncle took him to the Malayan peninsula, ostensibly to be one of three native guides on an expedition to New Guinea. But the trip was canceled, the "Director" asking instead if he would stay on to work in Gunong Tahan. The "Director" was Herbert Robinson, director of Museums, and founder of the MMS. Robinson asked Kedit to help build the meteorological station at Gunong Tahan, where he worked with four men Robinson had brought in from Europe as meteorological trainees. Here he learned how to be a meteorological assistant, making observations and fixing instruments. When the Gunong Tahan station closed in 1923, Kedit spent several years with one remaining European (the other three left the country) and helped build two new stations in Pahang, two in Kuala Lumpur, two in Perak. They refurbished Singapore's Mount Faber station and helped construct the forecasting unit at Kallang airbase. His career thus followed the same path as the meteorological service, moving from the Museums Department to the Survey Department before evolving into the MMS. Kedit was a meteorological assistant when he was photographed with Kelliher and the rest of the team in 1936, but he was promoted later to a technical assistant. In this later role, he no longer undertook observations but was placed in charge of instruments, traveling across the country every three months to inspect and repair all the meteorological equipment at the small stations.[69]

There are few details on the assistants or observers who manned these stations in the small towns and remote parts of the peninsula. We do know that at the same time of the photo, one Mr. Raja Gopal was transferred to the Kuala Lipis station and Mr. Moborak Ahmad, known by reputation as "the all-round sportsman of Kuala Lumpur," took over Gopal's post at remote Alor Star where he must have found his sporting opportunities more limited. The following year, Mr. S. Lazar of the Perak based Sitiawan station proceeded on transfer to the more southerly station at Fraser's Hill and was succeeded by Mr. Chong Wai from Singapore HQ where he would have joined Mr. Yeo Woon Cheong.[70] The reason for mentioning these moves is to engender some sense of the life of a meteorological observer. Shifting around was part and parcel of the job and not one that many people enjoyed, especially if they had families, as it entailed enforced periods at different stations around the country for the first five years after joining the service. It was therefore hard for the observers to maintain any kind of continuity in one place and many of the stations were remote and cut off from even small-town amenities.[71] The men who joined up therefore were often very young and would spend several years away from home before they could settle down. Their arduous role was placed at the level of a service subordinate, but it was one that was fundamental to the functioning of the service and to providing local forecasting by keeping the network of observations alive.

The more senior scientific officers, during the interwar years, joined the global scientific community. Malaya had its first in-person representation (instead of being represented by India) at international conferences, which included the meetings of the IMO (the forerunner to the World Meteorological Organisation), the Conference of Empire Meteorologists, and the Conferences of the Regional Meteorological Commission for the Far East, first held in Hong Kong in 1937.[72] These regional conferences would advance regional meteorological research and matters affecting forecasting for aviation, including the urgent matter of adopting a uniform code for storm warning signals across Asia and improving international cooperation on everyday weather reporting.[73] It was essential that Malayan interests be embedded into Asian and international interests as cooperation was key to protecting aviation. C. D. Stewart spent a great deal of his time overseas attending such meetings as a result, paradoxically facilitated by some of the same flight routes that were now under discussion.[74]

The meetings were often fraught with contention over subjects that might be considered most basic, however. The question of when weather observations should be made, for example, became mired in politics and nationalism. At the first Asian region meeting of the IMO in 1937, calls for all observations to be made simultaneously according to Greenwich Mean Time (GMT) were rebuffed, as this time zone was neither practical nor applicable in the Far East.[75] Stewart along with Charles Jeffries, director of the Hong Kong Observatory led proposals to adopt a new system where observations would be made four times per day, within a latitude of one hour. This was intended to strike a compromise between those who favored an Asian time zone and those who insisted on adopting strict GMT timings, a politically astute move but not necessarily popular. A resolution was finally proposed that observations should be made at local times, but with more observations added. This meant that, regardless of what time an observation was made in local context, there would always be a steady stream of observations available for comparison—especially for aircraft flying at night when observations were less frequently registered, as many stations were not manned twenty-four hours.

There were problems in submitting real-time information, however, especially in the tropical and remote areas of Asia and despite the new wireless technologies. Although the technology existed, the infrastructure to support it was not always in place. Many stations were ill-equipped and in isolated areas. Workarounds had to be instituted whereby observations could be transmitted across smaller areas in a form of relay and then onward to towns or cities where suitable telegraphy equipment was available to transmit the observations more widely.[76] These problems were compounded further by the issue of uniformity of observation and synoptic chart types then in use within Asia, with Stewart noting how Malaya alone used at least three different forms of chart.[77]

Despite these difficulties, there were some areas of concern that found unanimous support. These included the utmost importance of making nighttime and Sunday observations, that upper air temperature and humidity observations be made a priority, that all observations be transmitted only by wireless, and that new terrestrial stations equipped with the full power to transmit across distances be established to improve coverage.[78] A permanent subcommittee was put forward to manage the many ongoing difficulties of international communications and timetabling of synoptic issues.

Notwithstanding all the struggles, such meetings helped to create what Anduaga terms a "consensus atmosphere" across the first half of the twentieth century, one that helped to usher in acceptance of internationalization and uniformity in practice.[79] As Étienne Bruzon of the Phu-Lien Observatory in Vietnam acknowledged at the 1937 meeting, "great steps" had been made by all members in the spirit of "good will and co-operation."[80] By the late 1930s, the tone was upbeat with a new International Commission for Aeronautical Meteorology (1935) and discussions underway at the IMO as to whether there should be an intergovernmental organization with all national weather service directors as members. Led by colleagues in France and Norway, this initiative saw the IMO drafting a world meteorological convention to secure this intergovernmental status that was presented at the 1939 meeting of the International Meteorological Committee in Berlin. Unfortunately, the Second World War intervened.[81]

MALAYAN METEOROLOGY AT WAR

While Europe became mired in war, in Malaya during 1939 and 1940 most were doing business as usual. The wartime demand for rubber helped the colony revive its rubber industry by 1940 above and beyond its pre-slump levels and the war gave a major boost to the tin industry also.[82] In 1940, Singapore alone was valued at around £131.25 million to the British government.[83] For the most part, daily life in Malaya carried on as normal, although there was an unsurprising amount of tension between the Malayan people and the Japanese expatriate communities and, of course, fear for friends and family in other countries.[84] During these first months of normalcy, the MMS was—for reasons of national and imperial security, and like many other government departments—put on a war footing. Stewart did not see the war as director, as he had retired in December 1938, but his replacement, Hercules Moorhead, previously the head meteorologist in Bermuda, was launched right into the thick of it.[85] All senior MMS staff were redesignated into the Malayan Volunteer Air Force, the staff working at the Kallang Forecasting Office, for example, were mobilized under Mr. A. Grimes, the peacetime head of forecasting and now wartime wing commander at Air Headquarters Far East, working in confidential forecasting for the RAF.[86] Essentially, the MMS became the forecasting service for the military, a fact exemplified during a 1940 broadcast on Radio Malaya by Moorhead, where he described their new role of supporting military operations in the event of active warfare.[87]

Across 1940–1941, the war came closer to Malaya. The Japanese Army won a series of strategic victories in Vietnam, Cambodia, and finally, British Burma at the end of 1941. Intelligence reports suggested an impending Japanese plan to use Thailand as a launch base for a terrestrial incursion into Malaya and the next part of their campaign.[88] The defense of the northern Malay peninsula thus became a priority.[89] There was an RAF aerodrome at Kota Bharu, a small town on the east coast bordering Thailand. Orders were for the MMS to establish a second forecast office here in the north of the country to provide the RAF with tactical forecasting.

Britain and her allies were underprepared for the Japanese advance. The expectation had been for a series of air strikes, hence the defensive strategy of supplying the airbase as the key means of defense. The Japanese focused instead on a stealth strategy, moving over land through the dense Malayan jungle and mountainous terrain that cloaked visibility from the air.[90] The meteorological staff at the Kota Bharu airbase were therefore the first strategic base in the direct line of fire, to be attacked by invading Japanese forces moving south from Thailand on 8 December 1941. The invading forces continued a steady southward advance, and by February 1942, most of the key towns on the peninsula had been captured. All meteorological stations on the Malayan mainland were evacuated and their staff ordered to carry out a scorched earth policy, destroying instruments and records that could give the Japanese any advantage. In the fighting, at least eight meteorological staff were killed as they issued weather telegrams for the RAF's antiaircraft operations until the last possible moment.[91]

Britain surrendered to Japan on 15 February 1942. Over the next three years of occupation, the Malayan meteorological facilities were taken over by Japan's military until their own surrender on 5 September 1945. The new Japanese Air Force headquarters was based at Siglap and their meteorological units at the Kallang airfield. Under the command of Adjutant Kunitaro Ueda, the Kallang airfield became the chief forecasting and weather operations base for the Japanese Air Force for all Southeast Asia. All Japanese meteorological officers were sent from Japan to Singapore to be trained in tropical specificities, and from there they were sent out to work in Myanmar, Indonesia, Cambodia, and Vietnam in teams of fifteen or twenty people. Ueda later recalled how many of these men never came back but simply disappeared without trace into the jungles and remote areas of the region.[92]

The takeover and continuation of Malaya's scientific facilities has been commented on by contemporaries and by historians, signaling how science continued to operate and continued to be prioritized despite the change of regime and key officials. Indeed, in many cases, scientific personnel did not change.[93] The new authorities needed the assistance and advice of key employees and, importantly, the continuity of local staff in subordinate positions—such as meteorological observers at remote outstations—to enable the service to continue to run. Many of the Chinese, Indian, and Malay meteorological assistants and observers continued to work alongside the Japanese.

One of the local men who continued to work under the Japanese was Kanagasabai s/o Narayasamy, a fresh recruit at the time of the Japanese takeover. He worked at headquarters in the Fullerton building. His story might be typical of the many subordinate staff who continued to work in occupied Syonan-to, as Singapore was renamed. The new Japanese governors quickly sent Kanagasabai to man a meteorological station in the Cameron Highlands in Perak where he spent nine months making observations and generating weather reports. He worked shifts, as the station had to be manned twenty-four hours seven days a week, probably little different from the conditions under the British. He was contracted to work at this remote station for two years, but upon hearing that his father, then living in Singapore, was sick, he quit the job. He took on a position at the Naval Base because their food rations were better, and he was able to take these home to his family as he was based nearby. He returned to meteorological work under the British in 1945.[94] The British and Europeans officers who remained in Malaya were captured and interned for the duration of the war. Director Moorhead is an example. The details of his three years under the occupation are vague, but we know from his postwar "Liberation Questionnaire" that he spent much of this time as a prisoner at Sumatran internment camps. He survived, as did Grimes of the Kallang Forecasting Office, with both of them returning to their posts after 1945.[95]

When the Japanese surrendered in 1945, the MMS was in a mess. Despite the efforts of Ueda, vital instruments were missing, the service was underfunded and understaffed, and operations continued to be directed to the military needs of the British government, returning as the British Military Administration from 1945 to 1946. Compared to the pre-1929 days, however, Malayan meteorology, now with a dedicated department and a clear purpose, was in a far better place than it had been a generation earlier. Meteorology had risen to a occupy a new and essential scientific space, one that was and would continue to be intimately connected with the military and with aviation for many years to come.

PART II

THINKING WITH WEATHER

Trade, commerce, agriculture, and industry, these were the central tenets of the colonial enterprise. To the EEIC and, later, the British colonial government, the riches of the Malayan environment were fundamental incentives to settle on the Malayan peninsula. From here, the EEIC and Britain could hold a key strategic position in the East Indies spice trade and, as the world shifted, grow wealth from the forests and the commercial plantations. The region's tropical abundance and perpetual verdure were not due to the soil, argued John Cameron, then editor of the *Straits Times*, the region's English-language newspaper, in 1865, but to the climate. To the atmosphere was credited "the luxuriance and variety of the island's natural products" and the "daily occurrence of refreshing showers, and to the heavy night dews" that produced "green grass, the rich foliage, and rare fruits."[1]

The tropical climate underpinned Britain's political and economic reasoning for settling the Malayan peninsula, but, once certain key facts had been determined about the local atmosphere, the authorities ceased to be especially interested in the weather at a professional or organizational level. Officials turned their gaze to sciences that directly related to productivity, such as botany, horticulture, and agriculture, and even then, it was left to the ambitions and interests of prominent colonial gentlemen to form their own clubs, societies, and organizations for much of the nineteenth century. The situation of the Singapore Botanic Gardens is a case in point.

Sponsored by membership donation, the original "economic garden" as envisioned by Stamford Raffles had failed by 1829, and it took the combined efforts of the Agri-Horticultural Society—a group of prominent citizens and plantation owners—and a major donation from landowner Ho Ah Kay for it to get restarted in 1859.[2] As Timothy P. Barnard has argued, the Gardens might well have stalled again in the later nineteenth century, had it not been for the efforts and sheer willpower of the director Henry Ridley to keep botanical research on the government agenda.[3]

The story of botanical science in Malaya mirrors in some respects the story of weather science. Despite both sciences being critical in the British government's pursuit of economic productivity, for much of the nineteenth century the authorities in each case preferred to sponsor the end product rather than the inputs required to secure their goals. This changed in the early twentieth century with the advent of a new dedicated division for meteorology, which employed globally trained experts. Nonetheless, British support for the sciences in Malaya was still limited, as the explication in the chapters ahead suggest.

This part of the book moves on from exploring the colonial contexts of the development of weather science to examine instead the circumstances of the local climate that led to the interests and preoccupations of the colonial government and scientific establishment in Malaya. It argues, contrary to some scholars' views, that the environment within which science was practiced had a significant influence on the agenda and outcomes of scientific practitioners.[4] Still, it develops some of the theories proposed in the first part, that weather science was a lost opportunity in Malaya, suggesting some of the reasons why this might have been and, showing where imperial interests lay. It draws a direct correlation between the weather and the direction of science.

CHAPTER 3

THE GREAT DESICCATION DEBATE

GLOBAL KNOWLEDGE AND LOCAL CONTEXTS

The sciences of the atmosphere, in the sense that we understand them today, germinated across the nineteenth and early twentieth centuries. In the early days, it would not be appropriate to talk about climate science or atmospheric science, rather meteorology (as observation of the weather) and the interlinked organic sciences of air, water, and plants were the dominant schools of thought. Connecting across the schools was the theory of desiccation, which concerned changes to localized climates, especially the diminution of rainfall as caused by deforestation. Although desiccation theory was by no means without its detractors, by the 1880s the theory was widespread, exerting significant influence on the public, policymakers, and scientists, many of them convinced that deforestation was behind many of the major "natural" disasters of the time, especially floods and droughts.

The ideas underpinning desiccation had been in circulation for centuries. The classical thinker Theophrastus, for instance, posed questions concerning man's ability to exact climatic changes or to spur human-induced disasters, citing examples of small-scale temperature and rainfall variations due to woodland clearances in Larisa and Philippi.[1] In Ming Dynasty China, the scholar Yen Seng-Fang observed a connection between deforestation and a heightened propensity for flooding, as torrential rains ran unchecked down denuded mountainsides into plains and valleys, swelling rivers, breaking embankments, and devastating farmlands.[2] Nevertheless, it was to be the nineteenth century before desiccationist thinking became mainstream, peaking between the 1840s and 1890s. In part this upwelling can be attributed to better and faster systems of communication, alongside an upsurge in printed scientific materials such as journals, articles, and lectures that were available to a wide reading public. As Stephen Legg points out, knowledge about the perceived effects of human activity on localized climates became widespread during the nineteenth century thanks to the steam postal service and development of commercial news agencies.[3] It is no coincidence either that the same century witnessed an extension of commercial land development across many regions including America, Canada, Australia, China, and Southeast Asia. Colonialism too, especially in Africa and tropical Asia, created a situation ripe for incubating and disseminating desiccation theory as the rapid development of natural landscapes for agriculture, industry, and urbanization prompted many to fear irreversible environmental damage. As Richard Grove argued, "fear of artificially caused climate change, and much of modern conservation thinking which that anxiety stimulated, developed specifically in the tropical colonial context."[4] Other scholars have modified Grove's ideas on colonialism as a watershed in environmental thought, but it cannot be denied that the pursuit of commerce by imperial powers had a massive impact on land use in developing countries and that many contemporaries perceived this as a serious problem.[5]

The science of desiccation evolved, like many types of natural science at that time, from firsthand observational analysis. Alexander von Humboldt, one of the most famous natural scientists of the period and one of the first to propel the evils of deforestation into the limelight, argued based on his own experiences that the rapid destruction of forests in Venezuela had affected rainfall, streamflow, and lake water levels.[6] Humboldt's philosophy on nature was most famously espoused in *Cosmos* (published in five volumes from 1845 to 1862). Here, he vividly depicted the universe as a vast interlinked and balanced system, thus impact on one area could have repercussions elsewhere.[7] The sense of balance on the global—even universal—scale that Humboldt introduced became a leitmotif for the science of the age. François-Antoine Rauch, for instance, a French civil engineer, wrote of the "natural economy of water," an essentially limited cycle of evaporation, condensation, and rain where if one element was disturbed, say by the destruction of forest, there would be catastrophic results.[8]

These perspectives, or variations on the theme, were to strongly influence scholars working in the then developing fields of organic chemistry, or agricultural and forest sciences, who were attempting to understand linkages between atmospheric moisture, precipitation, soil quality, and trees. Jean Baptiste Boussingault, for instance, one of Humboldt's students, took the view that forests encouraged water retention in the surrounding soil, which was then released slowly over long periods. The extensive loss of trees would therefore result in a higher probability of flooding and flood-related soil erosion.[9] Boussingault's retention and evaporation research was to influence scientific forestry across Europe, America, and the European colonies.

So would the work of French scientist Antoine Becquerel, whose theories on forests, water, and heat were reproduced across popular and scholarly mediums globally for several decades.[10] His sweeping survey of deforestation in different countries divulged how specific geographies and climates, along with the extent and type of woodlands and soils, would all impact on water retention and humidity. He concluded that there could be no one-glove-fits-all theory. He did not definitively conclude that forests affected rainfall (though he was often cited as suggesting that they did), but he did note, similar to Boussingault, that trees affected running water, which then influenced erosion or stream and lake levels.[11] For Becquerel, the real climatic importance of forests was in their relationship with local thermal processes, in the cooling or raising of the temperature of surrounding air.

By mid-century, desiccationists were not lone voices but influencers of the environmental policies considered or enacted by leading scientific organizations of the day. In 1852, for example, a study was commissioned by the BAAS to investigate Humboldt's claim that "By felling trees . . . men in every climate prepare at once two calamities for future generations."[12] Noting that although it was currently hard to draw any definite conclusions because of the lack of weather observations from periods and areas prior to, and after, forest clearances in many places, the study concluded that there was nonetheless "little doubt that the entire removal of wood leads to the diminution of water," "atmospheric and surface moisture," soil exhaustion, and an increase in heat. The loss of woodland was also the loss of valuable building material, and the regrowth of "rank" and "unhealthy" scrubland in its place could even impact on human health.[13] As Brett M. Bennett and others have argued, it is clear that such studies were not solely motivated by conservationist principles but by the need to conserve and to exploit, one of the principle paradigms of late nineteenth and early twentieth century forestry.[14] The two—occasionally conflicting—principles worked to create a narrative that appealed to governments, scientists, and the public on many different levels and were not necessarily unaligned, as the preservation of forestland was also an essential part of effective economic forest management. Few could debate that forests did not need protecting, even if they differed in their reasons for so doing.

Building on this emerging narrative, the man often credited for bringing desiccation theory into the mainstream was George Perkins Marsh.[15] His pessimistic account of humanity's actions and the future of nature if the former were left unchecked were expounded in *The Study of Nature* (1860) and *Man and Nature* (1864). Inspired by Humboldtian science, Marsh also wrote extensively on the subject of balance. He explained how man's greed for natural resources had led to the extinction of many plant and animal species and the destabilization of natural ecosystems. On forests, he showed the role of trees in maintaining a perpetual water cycle. Trees contributed to humidity levels, kept local temperatures low, and helped maintain soil moisture and adequate rainfall levels, thus regulating water flow. This in turn kept the trees healthy, so their role as carbon absorbers achieved "perpetual" atmospheric "change."[16] The worst culprit for destabilizing this system was, in Marsh's opinion, the adaption of land for agriculture on a mass scale (which included deforestation), a trend that, although having historical precedent, had intensified in his modern age.[17] "Throughout the Eastern continent," Marsh thundered in 1860, "man has everywhere left his visible impress on the face of the material creation. He has stripped the mountainsides of their natural vegetable ornament and protection, and laid bare their surface to the influence of sun and wind and frost and rain . . . [over] thousands of years of assiduous toil with plough and harrow and spade and hoe."[18] This action had exacted "momentous consequences" to soil drainage, resulting in the desertification of once lush and fecund lands and extirpating vegetable and animal species.[19]

Marsh was not a lone voice. His writing engaged with a surprisingly loud and most definitely globally linked conservation narrative, particularly prevalent in America, Europe, and the European colonies in Asia. In a paper read at the 1877 meeting of the American Public Health Association in Chicago, for instance, Dr. George L. Andrew expounded on man's monumental responsibility for wanton environmental destruction, bemoaning humanity for "turning paradise into desert" and man's ability to "waste and destroy."[20] Likewise, referring to a recent British parliamentary paper, the *Pall Mall Gazette* reported in 1878 how it was essential that "prompt and comprehensive action" was taken to "stay the influences at work to destroy the indigenous forests" in the colonies. Noting "Dr Hooker's" views (presumably Joseph Hooker) on the subject for Ceylon, the author went on to write that "'little short of a meteorological revolution' was necessary to stem the destruction . . . streams once regarded as perennial now run dry under exposure to a tropical sun, and the periodicity of the rainfall has been seriously disturbed."[21]

Particularly significant in the British colonial context was the 1877 publication of *Forests and Moisture*, an epic treatise covering all aspects of the relationship between soil, hydrology, and forests by John Croumbie Brown, who had been professor of botany in Cape Town.[22] Brown was employed as colonial botanist at the Cape of Good Hope from 1862 to 1866 and was well-known for his work on hydrology and forest conservation in that region. As Fellow of the Royal Geographical Society and the Linnaean Society, he was well placed to become a vocal critic of deforestation and, it has been argued, a key voice in cementing the desiccation discourse into mainstream and popular science.[23] In *Forests and Moisture* he critiqued the prior studies by Marsh, Becquerel, Humboldt, Boussingault, and others on desiccation across different regions and chronologies and engaged with contemporary theories on atmospheric precipitation. Published at the peak of interest in desiccation, the text offered a simple, comprehensive, and compelling case for the influence of trees on climate, which was to prove a major inspiration to British forestry practices across the colonies. So strong was his belief in desiccation that Brown rode roughshod across potentially contradictory evidence, arguing that "it appears to be established as a fact that there are cases in which an extensive destruction of forests has been followed by a marked desiccation of soil and aridity of climate, and some cases in which the replanting of trees has been followed by a more or less complete restoration of humidity . . . the effects of forests in retarding the flow of the rainfall after its precipitation has been established, I consider, beyond all question, and not less so their effect in maintaining a general humidity of atmosphere and of soil."[24] Just as Becquerel and Marsh noted, Brown also pointed to trees' role in lowering temperatures in their immediate vicinity, though the relationship of trees with atmospheric moisture was, for him, the most critical.

The 1870s stand out in many respects as a critical decade of heightened discussions on this topic across the world. As Deborah Coen has noted for the Austrian and German context, the decade saw the publication of many influential texts and also meetings, such as Ernst Ebermayer's 1873 study of Bavarian forests or the International Statistical Congress held in Budapest in 1876 where scientists considered proposals for further study into the relationship between forests and climate. Although immediate or definite outcomes of any of these discussions were ambiguous, their existence reveals a shift toward thinking more about climate dynamically and across wider scales and a vigorous interest in forestry as a means to protect climate.[25]

DESICCATION AND HEALTH

Where some of these works differed was on the relationship between forests, climate, and humans. Brown, for instance, was writing extensively on transpiration and the science of aqueous vapors such as moist forest air and forest fogs and did not recognize or find of interest the link with prevailing medical ideas on miasma: vapor dangerous to human health and the role of trees in mitigating its effects. Marsh, though ostensibly writing from a purely environmental perspective, also proposed that, at the local level, trees absorbed miasmas considered injurious to human health. This feat had particular importance in marshy, boggy areas where rotting vegetable matter was thought to generate unwholesome airs.[26] The connection of forests, or trees, with health was widely recognized, perhaps because of the nineteenth-century fascination with medical climatology, which considered climatic influences on human health.

Bernhard E. Fernow, chief of the Forestry Division of the United States Department of Agriculture, was emblematic of the times. A vehement believer in the climatic influence of forests and the ability of trees to absorb toxic gases, he argued that they filtered fungi and bacteria from the air and contributed toward the production of soil that was rich in nitrates, phosphates, and potash in which unhealthy microbes were less able to thrive. Trees therefore played a crucial part in keeping diseases of miasmatic origin at bay.[27] Fernow was strongly influenced by the work of the "Father of Forest Meteorology," Ernst Ebermayer, especially his studies of soil and air moisture.[28]

Many medical doctors pursued the same philosophies. James M. Anders, a Philadelphia-based physician with a keen interest in the natural sciences, conducted a series of experiments in the 1880s. His plan was to determine the amount of water vaporized from known areas of leaf surface, land surface, and water with the intention of uncovering more about the relationship of forests with atmospheric temperature and rainfall.[29] Climatic changes caused by the destruction of forests, he later argued in *Sanitary Influences of Forest Growth*, had a significant impact on human health.[30] Anders's work was amply reproduced in the British colonial journals circulated among agriculturalists and administrators in India and the Straits Settlements.[31]

Into the twentieth century, in 1916 Henry Reed Hopkins, once president of the New York Medical Society, continued to argue the vital importance of forests to the American public. Clearly influenced by the contemporaneous outpouring of literature on the interaction of forest and climate, he further argued that this ecosystem had an even more vital role: maintaining human wellbeing. His opinion was based on the purported role of forests in the "production" of climate (temperature, air, light, and humidity) in creating an ideal, healthy environment for man.[32] "Drive your patient into the Black Forest with a thick stick and, if he comes out, drive him back again" was the opening quote of Henry M. Fisher's note on the health-giving properties of forest airs in 1886.[33]

In the colonial context, denudation was often thought to have increased malaria, and plantations were considered infinitely less healthy than original forest.[34] In Mauritius, for example, when Chief Medical Officer Francis Reid analysed the malaria epidemic of 1866–1867, he wrongly attributed it (though correctly according to prevalent theory of the day) to plant spores that had increased in the changed climate following forest clearances. Cleared land did not drain so effectively, he argued, rivers had suffered from declining rainfall, and the climate had become warmer, thus creating the ideal environmental conditions for the propagation of the disease.[35] George

Andrew, writing a few years later, came to the same conclusions, but on a global scale, asserting that forests influenced health by "their effects on atmospheric constituents, water supply, climate, and malaria."[36] The "breeding" of "dangerous miasma" at British colonies was also raised and critiqued in the popular press, as well as in the scientific literature.[37]

The association of miasma with bad health has been deeply ingrained into cultures across the world from time out of mind, but opinions on how these are associated have differed widely and often offered contradictory "evidence" of forests' part in creating a healthful (or not) environment. In seventeenth-century England John Evelyn, the notable campaigner for forest preservation, wrote that the felling of trees might have a beneficial effect on the health by "letting in the air and sun . . . so that these gloomy tracts are now healthy and habitable."[38] The idea that tropical forests, because of their humidity, were breeding grounds for disease provided one solid justification for the extensive forest clearances in the early British Caribbean colonies. That tropical forest clearances might be beneficial to improving health could only have helped support economic deforestation and land policies under the British colonial regime.

Indeed, in 1875, Franklin B. Hough pointed out that, although disease (in this instance malaria, originally considered a miasmatic complaint) increased immediately following forest clearance, when land was properly drained, the cleared land became healthier. In his opinion, the humidity of the forest held numerous dangers: "In a region thus densely covered, the soil and the stagnant air near the surface, are at all seasons overcharged and often saturated with moisture, and the decaying vegetation always present, will, in hot climates and seasons, tend oftentimes to malarious diseases of a typhoid, and sometimes of a malignant type . . . in most cases, where the drainage can be made complete, these conditions will wear away, as the organic matter in the soil becomes fully decomposed; and a region once noted for insalubrity may become quite free from malarial diseases, and even exceptionally healthy."[39]

Hugh Williamson, physician, and a voice for the benefits of forest clearance, also identified deforestation in combination with drainage as the best method to achieve a healthful environment. In his opinion, removing trees would create a more temperate climate in which "we shall seldom be visited by frosts or snows" and the removal of tall, dense woodland would allow freer circulation of air, especially fresh breezes off the sea. Prior to land development, he went on to add, the air "was constantly charged with a gross putrescent fluid" caused by the presence of marshes and the "copious perspiration" of plants.[40]

Miasma theory was at a critical stage of evolution in the mid to late nineteenth century; after centuries of consistent belief in the dangers of noxious smells, finally there was evidence of germ theory and of waterborne diseases by century end. Much of this came from studies of epidemics such as cholera that affected temperate and tropical environments alike, but the specific nature of forest air was a subject unto itself.[41] The whole body of work devoted to desiccation and deforestation was full of contradiction and confusion. Many influential scientists cum meteorologists sat on the fence, such as Robert Ellery, the government astronomer in Victoria, Australia, or American US Weather Bureau's Aldophus Greely, suggesting evidence that both supported and contradicted deforestation-induced aridity. Henry Russell, another notable government astronomer and meteorologist (in Sydney), was perhaps clearer, suggesting that a lunar-influenced cycle was more important in creating drought and flood situations.[42]

In 1904 Peter Kropotkin, more commonly known for his anarchistic political work, also delivered an account of the desiccation in Central Asia.[43] Citing historical examples gathered by the Austrian geographer and explorer Oscar Lenz, Kropotkin argued that the arid regions of Turkestan and Central Mongolia had not always been so. In Siberia and "European Russia" too, "immense areas, formerly under lakes and marshes, have been laid dry." Kropotkin noted that, although blame for this change had been placed squarely at the door of three centuries of deforestation, instrumental weather observation showed that tree loss (massive though it was) could not adequately explain the scale of desiccation in his region of study. This, he explained, was due to the natural geological fluctuation of the postglacial period, another idea on macro-scale climatic changes that had been taking hold across the nineteenth century after the pioneering work of Louis Agassiz.[44]

Despite the existence of counternarratives, desiccation became a popular explanation for perceived climatic fluctuations in colonial officialdom, especially in regions with a prevalence of unpredictable drought. Speaking at a meeting of the Indian Section of the Society of Arts in 1881, for instance, the former governor of Bombay Presidency Sir Richard Temple argued that extensive regional deforestation may have contributed to the droughts during his four decades serving in India until 1880.[45] So too did John Augustus Voelcker, consulting chemist for the British Royal Agricultural Society of England. In this role, Voelcker undertook a major investigation into Indian agricultural improvement. In his report, he noted how the rainfall of Udhagamandalam (Ootacamund) station in Tamil Nadu increased over a period of twenty years between 1870 and 1890 in tandem with reforestation in the area. He concluded, albeit not definitively, that forests could favorably influence the climate over a generation.[46] British imperial forestry was at its most developed in India, a country where conservation narratives developed in response to extremes of weather, soil erosion, and the necessity of preserving valuable timber resources, Indian forestry becoming a model for the rest of the British Asian empire, especially Burma and the Straits Settlements.[47]

GLOBAL DESICCATION NARRATIVES IN LOCAL CONTEXT

Deforestation was a problem in the Straits Settlements, especially in Singapore and in Penang, where, for instance, as early as 1830 a large part of the "level ground" and valley was converted to growing rice, pepper, and spice, dramatically changing the natural landscape.[48] Many hills were cleared to make way for nutmeg and clove plantations, criss-crossed with new roads and houses in "place of the useless forest which once covered them."[49] This resulted in the formation of dangerous unhealthy miasmatic airs across the eastern hill ranges, stated Dr. T. M. Ward of the Madras Army, then stationed in Malaya. In his opinion it would take at least three years for the miasma to dissipate from the newly cleared earth, so he advised extra precautions—such as avoiding night dews—to protect settlers' health.[50] Gambier and tapioca were also cultivated extensively in the early years of the settlements. Both were considered the scourge of the forests, necessitating large-scale clearances, which leeched the soil to the point that it became exhausted and worthless within short periods. Gambier—described as the "locust of cultivation"—may well have resulted in the loss and abandonment of up to forty-five thousand acres of Penang Island by 1859.[51] H. Price, writing on the early history of planting in Singapore, claimed that tapioca cultivation laid waste to the soil for up to fifteen years, in which time "the abandoned tapioca estates turned into . . . extensive stretches of *lalang*," a particularly tough and virulent grass weed that proved troublesome to remove.[52]

By the 1830s and 1840s the extent (and cost) of land-use change in the Straits Settlements had become very apparent, mirroring in many respects the situation across British colonies globally. Citing the British experience in India, the Caribbean, and Mauritius especially, Richard Grove has argued that these two decades were critical in rasing awareness, a trend that contributed directly to the establishment of the Bombay Forest Department in 1847.[53] In the Malay Peninsula, the Indian discourse proliferated via early journals, the press, and the officers who had worked across different Asian stations in the course of their careers. One such man was the short-lived Thomas Newbold, originally of the Madras Light Infantry under the EEIC, whose posts took him between India and the Straits Settlements, the latter for three years from 1832. He was a contributor to the *Journal of the Straits Branch of the Royal Asiatic Society* and to the *Madras Journal of Literature and Science* and was later elected Fellow of the Royal Society.[54] Very well aware of the work of Boussingault, Newbold made a connection between deforestation and desertification during his extensive researches in India in the late 1830s, drawing from the local knowledge of indigenous inhabitants.[55] Traveling near Bellary (Bellari in modern Karnataka state) he was surprised to discover great extents of sand and sand dunes, seemingly out of place in the region. Talking to the local headman and the oldest inhabitants at Honoor (Honnur), a small village some twenty miles southeast from Bellary, Newbold discovered that the sand was progressing annually. The desertification accelerated during major drought periods, such as the villagers recalled in 1828. To diminish the speed at which the sands were threatening the settlement, villagers were allowing trees that previously were cut back to regrow, because they attributed the desertification to deforestation. Around five miles away were the ruins of another village that Newbold referred to as "Bodurti," which had been entirely subsumed by the dunes.[56]

Newbold was a well-known name on the British Asiatic journal circuit by the time James Logan came on the scene. Moving to Penang in 1839, Logan was not a military man like Newbold, but they shared an interest in the languages, natural history, and geology of the Straits Settlements and of India. In the *Journal of the Indian Archipelago and Eastern Asia* that he launched in 1847, Logan published articles by notable Europeans based in India and the Straits Settlements. For Singapore and Penang, for example, John Turnbull Thomson (government surveyor from 1841 to 1853), Dr. Thomas Oxley (senior surgeon), Lieutenant-Colonel James Low, and medical doctor Robert Little were all regular contributors who made forays into the desiccation debate.[57] Thomson, for example, published a detailed description of the geography and geology of Singapore in 1845. Noting that in many places on Singapore Island, such as Kallang, great quantities of spongy decayed vegetation assisted in retaining moisture, thus acting as a natural constraining mechanism in preventing large-scale flooding.[58] Low and Little both wrote about atmospheric moisture, ventilation, and health, and their work was disseminated in the popular press as well as in the special interest journals.[59]

Logan himself was a prolific contributor to his own journal and to later publications including the *Penang Gazette*. He was an early and outspoken critic of deforestation. Published in the second edition of the journal in 1848, his "Probable Effects on the Climate of Pinang of the Continued Destruction of its Hill Jungles" was a wake-up call to colonial administrators on the peninsula as elsewhere in Asia and quite plausibly the first major published article on this subject for the Straits Settlements. Logan argued that extensive jungle clearances were destabilizing the soils and climate of the hill slopes on Penang Island. As the root cause of the issue he singled out Chinese "squatters," who cleared trees for charcoal burning and then abandoned the area, after the soil became less fertile. Although referencing the Chinese cultivators as the cause, he also placed blame on the British government for a lack of foresight, regulation, and policy. In a statement not far removed from modern discourse on climate change, he maintained that "climate concerns the whole community, and its protection from injury is one of the duties of Government." Interestingly, he referenced unspecified hills in Singapore (possibly Bukit Timah) that had been protected by the "enlightened" governor there. Given that the gambier planting area on Singapore's hills was noted at around this time as a place of major deforestation, this seems a curious statement, but Logan may have been alluding to actions then being taken, as opposed to defending what had been done in the past.[60] In his paper, which was also presented in the form of a talk for the Asiatic Society of Bengal, he reveals contemporary perceptions of the desiccation science then known in the Straits Settlements, but more important, he references local actors taking on board contemporary thinking about climate change and how they were taking (small) steps to address it.

THE DEBATE BECOMES PUBLIC PROPERTY AND GAINS NOTORIETY

Nevertheless, these steps were not enough. Rampant deforestation continued to escalate as mining, plantation farming, and urbanization intensified across the peninsula over the next two decades. The discussion came to very public attention in 1865, in response to a damning report on forestry by Alexander Maingay, a surgeon and magistrate then stationed at Malacca. Like those of other colonial officials of the time, Maingay's professional life and personal interests were extensive. He had a great passion for botany—collecting plant specimens in Malaysia, China, and Burma—and in his official capacity was responsible for meteorology as well as medicine at the Malacca prison.[61] In 1865 he was commissioned by the government to compile a report on timber and forests for the Malacca settlement, which he completed that year. As Logan had suggested over a decade previously, Maingay also blamed shifting cultivators, especially those involved in tapioca and hill paddy farming, for denuding the forests in the region. "The ruin of the forest," he claimed, "also involves a corresponding diminution of the rain fall and a general elevation of the temperature, with an increasing prevalence of long droughts."[62]

The report unleashed an outpouring of opinion in the press over the following months, ranging from approval to denigration. "We are indeed glad to find him speaking out so boldly in defence of the preservation of our forests," the editor of the *Straits Times* wrote approvingly on 9 September 1865. "We have long protested against the gradual destruction of these," he continued, "and we earnestly recommend consideration to one of the concluding sentences in Dr. Maingay's letter in which he makes the following remarkable statement. 'I consider that the forests alone prevent the Malayan Peninsula from becoming an absolutely arid and relatively sterile, as Aden, the Ladrone Islands, or HongKong [sic]'."[63] On 14 September, an astute but less complimentary writer took up the narrative in the *Singapore Free Press and Mercantile Advertiser.* While commending Maingay as "a Government official in the Settlement who is alive to the importance of the subject," the anonymous author picked over Maingay's statements, suggesting that his conclusions actually countered evidence from Malacca's available weather observations. He then presents his own theory: that rainfall was mainly determined by geography. Situated near the sea, he argued, Malacca benefited from atmospheric vapor brought across from the Indian Ocean. The same applied to Singapore. Both would have been far wetter were it not for Sumatra's positioning, denying the west coast of the peninsula from many of the rains they might otherwise have enjoyed. This, in his opinion, was the explanation for the supposedly low rainfall. The author based this conclusion on his understanding of the monsoon (though he failed to explain monsoon failures or droughts—assigning these to unspecified "natural causes"). While he admits that there was some connection between deforestation and climate change, he argues that Maingay overstated the case. There had been only a "trifling" temperature or rainfall change since the time of Captain Elliot (the EEIC officer responsible for the Singapore Observatory readings from 1841 to 1845). This, he suggested, amounted to marginally lower nighttime temperatures and approximately one and a half less inches of rainfall per annum.[64]

On 28 September, the thread was once more taken up in the *Singapore Free Press and Mercantile Advertiser.* The anonymous author of this third response held a similar viewpoint to that of the item published on 14 September but was likely a different person. He presented his case with more reasoned and informed arguments than the former writer, drawing extensively from contemporary literature and methods. He also proposed an alternative to the desiccation or geographic explanations for declining rainfall. His key points were that the situation was appreciably more complex than Maingay allowed and that the evidence to support declining rainfall by deforestation, or indeed by any means, was inconclusive. Again, too, he also called for better weather observations, referencing the sentiments of James Logan in 1848, writing that "before we can safely reduce our experience into general laws, we must have regular contemporaneous observations carried on at our settlements for a period of not less than ten years."[65]

He also attributed this failure to government for not investing in the necessary instruments to make the observations. In particular, he referred to Jonas Daniel Vaughan, an estate owner who had volunteered and undertaken weather observations in "the interests" of the colony in 1856 but had "received no encouragement" from government.[66] Convinced of the importance of improved meteorological research, when he was on home leave in 1861, Vaughan personally approached Admiral Robert Fitzroy, the head of the newly established British Meteorological Service, to request weather instruments, which the colonial service had promised but failed to provide. Fitzroy was a willing supporter and enabled Vaughan to return with a new set of instruments. Not all the instruments made it to Singapore in one piece, however. Two pairs of thermometers, a rain gauge and "a somewhat injured" barometer survived the journey. The author then moves on to ponder an alternative theory to desiccation, that "the droughts are due chiefly to the direction of the wind in the Gulf of Siam, and China Sea; and partially to the decrease in the evaporation of those waters caused by the cold N.E. wind when the winter in the north is unusually severe."[67] Nevertheless, he rightly notes that the same criticism applies to his own work as much as Maingay's—neither theory can be tested without better meteorological datasets—adding fuel to his case for more registering stations and better records to be kept countrywide.

So, who were these authors and why did they feel justified in commenting on Maingay's report? Each piece was an editorial published anonymously, so, given the lack of any author identity and the limited budgets on which the papers operated at that time, there is a chance they may have been penned by the newspaper editors.[68] Certainly both contemporary editors were well placed to comment. The *Straits Times* had been sold to John Cameron in 1861. Formerly a master mariner, Cameron had a sailor's firsthand knowledge of the weather and, on board ship, would have had responsibility for maintaining the logbooks in which meteorological observations were documented.[69] Cameron's renowned work, *Our Tropical Possessions in Malayan India*, was published four years later, which tome devoted an entire chapter to the climate and environment of the region. On rainfall, he noted that it had latterly rained more frequently but less heavily in Singapore than before, a fact he ascribed to the clearance of trees from the town.[70] Could he have been the author of the first piece? The *Singapore Free Press*, with the less favorable editorials, was founded to represent the interests of the mercantile community, and its editor in 1865 was Jonas Daniel Vaughan.[71] Similar to Cameron, Vaughan was a ship's officer for the EEIC before landing in the Straits Settlements. He first worked as a police superintendent at Penang, before training in law and eventually becoming a judge in Singapore.[72] Responsible for keeping one of the earliest continual observational weather series for Singapore (facilitated by Fitzroy's instruments) between 1862 and 1869, he clearly had a deep and personal interest in meteorology, which, if the writer and editor were one and the same, might explain why the second article contained such a detailed call for more investment in the science.[73]

The difference in tone in the third article and the fact that the author wrote extensively about Vaughan's experiences in the third person strongly suggest yet another author. But this is all pure speculation. There are numerous candidates from around that time who may have had reason to contest the original report, all of whom had some experience and/or some knowledge of meteorology. J. J. L Wheatley, for example, was an apothecary and medical assistant who worked in Singapore and, later, in Johore for the colonial Medical Department. He was likely involved in coordinating the meteorological observations for those areas and was author of one of the few articles on local meteorological conditions published that century.[74] In an article published in 1881 he included an extensive review of contemporary literature on rainfall determinants and offered his own opinions on the causes of precipitation changes, chief of which were geography and the monsoon. Referring to an earlier spate of correspondence in the press about the matter and calling for better observational datasets, he included themes and arguments in his 1881 article that bear a striking resemblance to those presented in the newspaper item of 14 September 1865. Given the long time frame between the two publications, however, it is almost impossible to make any assumptions. Another possible author is Allan Maclean Skinner, who was colonial treasurer for the Straits Settlements. His reasons for being interested in the meteorology of the region are obscure, but he was keenly aware of the local discussion. An article he published in 1883 on meteorology in Malaya reveals that he was well-versed in the latest global developments in the field.[75]

A few months later, Maingay responded to his critics in an open letter to the editor of the *Straits Times*. Seeking to redress what the author of the article published on 14 September referred to as errors, he provided an in-depth account of the data and research upon which his conclusions were made.[76] Rich with references to the work of contemporary experts, using examples of deforestation and desertification from around the world, and presenting several years' worth of weather observations, he made a solid case. He also refuted his critic's claim that he had misunderstood the effects of the monsoon on Malacca and explained how the specific geography of the area worked to modify the monsoon's influence.

FOREST PROTECTION, COLONIAL RENAISSANCE MEN, AND METEOROLOGY

Maingay's reply seems to have ended this particular conversation, but the wider desiccation debate was only just beginning. Maingay was not alive to witness the next installment, he was shot and killed during a riot in Rangoon in 1869.[77] But the discussion continued in the press across the 1870s and was taken up by a series of prominent officers and governors as a policy concern.[78] In 1874, PCMO H. L. Randall made an explicit connection between the clearing of forests in Johore with a decline in rainfall from 1870 to 1873. The "only cause" in his opinion "to which this great decrease of Rainfall may be attributed" was "the extensive clearing of forests on the mainland," a state that could only be remedied by conserving large areas of forestland.[79] Only three years later, one of the strongest El Niño events of the nineteenth century occurred during 1877 and acted to spur talk into action. The event caused drought conditions, crop failures, illness, and hardship across the peninsula.[80] Contemporaries did not then know the atmospheric mechanisms that caused this event, so one of the ideas put forward by way of explanation for the extreme drought conditions was that deforestation induced climate change.

In 1879, Colonial Secretary Cecil Clementi Smith requested John F. A. McNair to conduct a survey of forests and timber reserves in the Settlements. McNair, then colonial engineer and surveyor-general, was responsible for some of the most extensive and well-known surveys and public works of the nineteenth-century Straits Settlements. McNair's report was an indictment of government policy (or lack thereof) on forests, which had resulted in indiscriminate and widespread wastage and quality timber shortages from illegal and legal felling and charcoal burning.[81] Logan had highlighted the charcoal burning as a serious problem back in 1848, yet the government did little in terms of regulation since. By the 1870s then, the problem was even more pressing. By way of example, two years prior to the release of McNair's report, Singapore's superintendent of the Botanic Gardens and government botanist, Henry Murton, had been engaged in a two-month research expedition in Perak, supported by Hugh Low, fourth Resident of Perak (1877–1889), ostensibly to explore local flora and economic crops there. Murton completed his assignment despite suffering from bad weather, a fever, sunstroke, and a twenty-foot fall.[82] Stationed at Gapis in the November, around two hours' walk west of Kuala Kangsar, Murton's team spent several days exploring the hills of Bukit Gantang and Gunong Bubu. He described the scenes trekking the slopes of Gunong Bubu (the peak of what is today Ulu Kenas Recreational Forest), dotted about with charcoal burners who were fast destroying the largest trees, and disclosed that "It seems almost like Vandalism to cut down all the fine Dipterocarps, &c., for the sole purpose of making charcoal, when the timber is so valuable, especially when smaller trees would produce equally good charcoal, but would perhaps not give such good returns to the burners."[83] Murton's dismay at the blatant forest destruction is a good example of colonial views on the role of the environment as an economic resource.

Resource preservation would also have been a strong motivating factor in Clementi Smith's interest in surveying forestland, but it was not the sole reason. He also requested that McNair investigate the "question of rainfall, and the relation that it bears to the retention or otherwise of our forest land."[84]

McNair's response to Clementi Smith's special request was hesitant, perhaps because he preferred to avoid claiming ownership of one side of the debate in what he termed "a point of so much general interest." He may have been party to the numerous conflicting articles in the press over the years, and he certainly would have been acquainted with some of the key speakers on the subject. The opinion he did offer was that the ocean and the wind played a larger role than the trees in determining rainfall, yet he did allow that denuded land would have a higher, faster evaporation rate because of the lack of forest cover to block the tropical sun. Forest reserves could mitigate this issue, he argued, and in his report he noted how a small reserve in each settlement was already allocated for this specific "climatic" purpose.[85] Declining further comment, he referred the whole question to PCMO T. Irvine Rowell—who at that point in time was responsible for meteorology in the FMS and the Straits Settlements—as the resident expert. Despite his recalcitrance in this issue, McNair's survey and report led to a new forest section being created within the Singapore Botanic Gardens jurisdiction in the 1880s, laying the groundwork for the wide-scale reservation of forestlands under its director Nathaniel Cantley, that decade.

The great drought of 1877–1878 was still in recent memory, and efforts to direct more attention to forestry and forest research began in earnest in the early 1880s. Under Nathaniel Cantley, who was director for the Singapore Botanic Gardens from 1880 to 1888, a further report on the colony's forests was completed. Forests had until this stage always been considered the purview of the Botanic Garden's administration, subsumed into their daily business and limited budget, largely because they were the chief scientific department then operating on the peninsula. Cantley's 1882 report marked a watershed in how the forests were viewed, no longer as an item point on the garden's annual report but as a subject worthy of a separate addenda. Cantley divided his report into two sections, the first about the current state of the forests and the second with his recommendations for improvement. His first point was that government had made no effort on the conservation front and, because of the reckless cultivation going on as a result of regulation and policing (mainly by the Chinese planters), timber supply was perilous and the climate was "sensibly affected."[86] The climate-altering effects of deforestation were given pole position as an argument for conservation, with Cantley arguing that, because of the unchecked destruction, showers had become less frequent and more localized and droughts of unprecedented length (referring to 1877–1878) had occurred. He also underlined the value of meteorological knowledge for foresters, for understanding not only weather for planting and germination but also atmospheric humidity, the crux of maintaining the "luxuriant vegetation" known to the region.[87]

At the same time, Cantley's connected all his climatic arguments to agricultural productivity, providing a great incentive to keep government interested. Singapore was experiencing altered rainfall patterns during dry seasons, he argued, with most of the rains falling only on the hills. This meant that temperatures on the plains had increased, and the many small streams relied upon for irrigation had dried up. Generally speaking, lowland agricultural areas were not receiving as much rain or streamflow as they had previously. He exemplified this point by presenting tables of rainfall with data collected over a period of twelve years at two different locations. The first was the criminal prison located right in the urban heart of Singapore, between Bras Basah and Stamford Road; the second was at Thomson Road, going out of town between Toa Payoh and the Waterworks or Impounding Reservoir (now McRitchie Reservoir) in a mix of rural plantation and forest. The latter rain gauge had recorded an average of thirty-four more rainy days a year for the period in question, which Cantley attributed directly to its higher, more forested environment, thus supporting his claim for trees' positive influence on rainfall.[88]

His investigations led him to conclude that, although average rainfall had decreased a little, the biggest change was to the number of days experiencing rain, as well as where it was falling. Relating to the future of agriculture and hydraulic management in the colony as well as to forests, Cantley revealed both the value of forests and the extent of investment necessary to preserve environmental and climate equilibrium for the future. He also did much to further the call put forward publicly since the 1860s for increasing the number of operational rain gauges across the peninsula. Fully resolving the question of denudation and climate change was, he claimed, of the utmost importance for the country.

By modern standards, Cantley's conclusions are full of holes. Based on observations of only two stations over a short period (climatically speaking), with limited understanding of natural decadal variability or, indeed, the El Niño Southern Oscillation (ENSO), they do not stand up to scrutiny. The strong El Niño of 1877–1878 was a one-in-a-century event, significantly impacting on the small dataset that Cantley used to make his report, making it hard to infer any past, present, or even future trends. If Cantley had had even a few more years' data to balance out the exceptionally dry weather of 1877, his deductions may well have been less categorical.[89]

Cantley's recommendations in 1882 were, for the most part, adopted. In Singapore, twelve reserves were laid out, totaling 11,554 acres, and staff were hired to manage them, under the initial direction of the Botanic Gardens. A map appended to the report shows exactly the extent of land to be reserved, a significant amount of which formed the basis of current reservation areas today.[90] PCMO T. Irvine Rowell was so impressed with the changes afoot that he claimed in his 1885 meteorological report that denudation in Singapore was "happily now checked."[91] Indeed, that year's annual government report suggested that monies had continued to be put aside for both reservation and the planting of new trees.[92] Irvine Rowell was wholeheartedly sold on the idea of desiccation. That "forest desiccation does influence rainfall materially, there can be little doubt," he also claimed in 1885. "Of this, many proofs now exist."[93]

In this opinion he joined other medical officers and colonial officials more generally who continued to write on the subject throughout the 1880s. Medical assistant J. J. L. Wheatley, for example, added to this dialogue in his article on rainfall published in the *Journal of the Straits Branch of the Royal Asiatic Society* in 1881.[94] His article was ostensibly written in response to the extensive ongoing public discussion on forest clearances and climatic changes. Alluding to the outpouring of letters on the subject in the local press, he stated, "how difficult it is to arrive at any definite conclusion" because of the limited number of observations available for study.[95] Having said that, Wheatley makes use of observations made as early as the 1820s and cites three datasets instead of Cantley's two, although he recognizes that the earlier sets were very patchy, that there was a lack of consistency in how the information was recorded, and that there were unexplained discrepancies between the datasets. The lack of concrete information, he argued, made it difficult to ascribe cause or effect in the debate. He put forward a solid case for investing more in local meteorological research, so that "in the perhaps not too distant future," all the influences of atmosphere and the internal and external "workings of this vast globe" will be "made clear, certain and simple; and the perils of the navigator at seas, the devastating effects of hurricanes on land, and the distress and want of famines will be foreseen and provided against with certainty."[96] Wheatley, like his contemporary Allan Maclean Skinner, also pointed to the possibility of other forces that might interact with the climate. Referring to contemporary science concerning sunspot activity and the solar cycle, he mused whether these might act as an external influence on annual rainfall levels. This was a field of popular interest since the development of sunspot photography in the 1870s and 1880s.[97]

Hugh Low was also an important voice urging more research into forests and climate in the region. Low—though ironically more often associated with encouraging tin mining, which resulted in localized environmental destruction—like many of his contemporaries fell into the mold of the colonial Renaissance man, dabbling in botany, forestry, exploration, agriculture, and horticulture.[98] He was a Fellow of the Linnean and Zoological Societies of London, and his knowledge of contemporary natural sciences was prolific. He took advantage of contemporary political interest in reforestation to introduce some of the first forest preservation policies in Perak with climate preservation in mind.[99]

Politically, this was a good time, because Lord Carnarvon, then British secretary to the colonies (1866–1867, 1874–1878) was advocating replanting as a means to avoid a disastrous resource crisis in deforested regions. Carnarvon was influenced by the work Joseph Hooker of Kew Botanic Gardens on deforestation and climate change.[100] Hooker in turn was influenced by the terrible loss of trees in Ceylon across the 1860s and 1870s, as well as by the earlier lesson of Ascension Island.[101] Working with Charles Darwin, Hooker encouraged reforestation on Ascension, strategically planting trees to manufacture a new ecosystem within two decades.[102] Correspondence between Hooker and Dietrich Brandis, inspector general of forests in India from 1864 to 1883, reveals their conversations on forests in tropical and subtropical settings.[103] Following Hooker at Kew was his son-in-law William Thistleton-Dyer who continued research on deforestation and its climatic impacts. As botanical advisor to the secretary of state for the colonies, Thistleton-Dyer was keen to push through preservation initiatives in all regions, including the Straits Settlements.

In this, however, he was hamstrung by the presence of William Edward Maxwell who was against any form of independent forestry department and who even wanted to take forestry away from the purview of the Botanic Gardens. Luckily, Resident General Frank Swettenham and Residents of forest-rich states William Treacher (a Resident of Perak) and Hugh Clifford (Resident at Pahang) recognized the economic benefits of a professional and centralized service.[104] A recent boom in the tin industry required quality timber for building new railways to transport the goods, hence the renewed importance of economic forestry. H. C. Hill, then inspector general of forests in the North-west Provinces and Oudh in India, was brought in to conduct a survey and to produce a report outlining an independent forest service for the region. His recommendations led directly to the formation of a new forestry service in 1901 and to the appointment of A. M. Burns-Murdoch as conservator of forests for the FMS and Straits Settlements.[105]

A LACK OF PROGRESS?

By the early twentieth century, despite the seeming progress in reserving forests, in performing forest and climate research, and in the creation of a dedicated Forest Department, the emphasis remained firmly on economic productivity rather than on conservation for its own sake. The public were aware of this discrepancy and were keen to blame government for official shortsightedness. In 1905, for instance, the local press published several items by outraged members of the public on the sale of land around Singapore. The discussion centered on what might today be termed the silo thinking of the different government departments. "Many of the older readers of the *Singapore Free Press* will remember the regular statistical articles on local rainfall so carefully compiled by the late J. D. Vaughan in the early eighties," wrote an anonymous contributor to that paper in June, continuing in wonderment as to what Vaughan might say

> were he to witness the immense spread of the very thing he so greatly dreaded, the denudation of our local jungle to make way for such an agricultural triviality as pineapple cultivation. As a result of this growing denudation we know that year by year our rainfall tends to become more and more irregular, even, in occasional years, precarious. . . . The Land Department, again, in its natural policy is entirely opposed to what we must assume to be the most important local function of the Forest Department . . . [they are] the one deadly enemy of our rainfall.[106]

The anonymous author of a follow-up piece published in November of the same year also noted how the revenue benefits of the Land Office sale of forestland around Singapore for pineapple cultivation would be nullified by the expenditure subsequently required to negate the effects of declining rainfall induced by denudation. Thus, the municipality and the ratepayer would be shouldered with the financial burden of improving water supplies and reservoirs if droughts increased. The anonymous author went on to argue that the government had a responsibility to manage this process more intelligently, to reforest the country and thus impose less strain on water supply.[107] Another letter on the matter published in March 1906 referenced the ill-effects of both rubber and pineapple cultivation on rainfall around the island. Referring to the "great afforestation question" in 1908, another contributor was to rain scorn on the many people who still "do not believe in the beneficial effects that forests exert on rainfall, although in most countries it is now recognised as a fact."[108]

In addition, despite the passage of more than five decades since the first calls for improving weather observations in order to resolve the debate, by the 1910s there was still criticism of the way the government was handling meteorology in the country, stating that the existing records were still not fit for purpose. Much of this criticism now came from the forestry community as well as from the medical establishment and the wide array of colonial officers. B. H. F. Barnard, at different points of his career a forest officer in the new Forest Department for Perak and Selangor, in 1911 published a scathing attack on the quality and scale of observations that underpinned research on the impacts of denudation. He deplored the fact that they had never been studied by a qualified meteorologist.[109] This is an important point and one supported by the evidence presented here; the writers on the subject of meteorology in the Straits Settlements were botanists, doctors, administrators, colonial administrators, and concerned members of the public. The great desiccation debate never again reached the levels of attention it received from the 1860s to the 1880s—it had a tendency to resurface in the popular and scientific press during and shortly after drought periods (1911 being once such example), but it ceases to appear in the Malayan context after the 1920s. Perhaps everyone hoped that the opening of the Malayan Meteorological Service (MMS) in 1929 would take the conversation out of the public sphere and into the hands of more expert researchers and investigators.

CHAPTER 4

HOT IN THE CITY

TROPICAL TEMPERATURES AND URBAN HEAT

In the 1930s Roland Braddell, autobiographer and long-term resident of Singapore, wrote that people who did not know Singapore well generally believed it to be hotter than it really was. The temperature, he argued, ranged between 76°F (24.4°C) and 87°F (30.5°C) in the shade, thus it always felt agreeable. It was the newcomers who were not yet acclimatized to the lack of variation and humidity who created an "exaggerated notion" of an uncomfortable climate and an unpleasant and unfriendly atmosphere. "Nowadays," Braddell wrote in reference to changing medical thought on tropical weathers, "the sun is regarded as a friend and not an enemy . . . and doctors are recommending people not to fear the sun . . . the real fact is that the climate gets blamed for all sorts of things which really boil down to over-exercise, over-eating, or over-drinking; and they are liable to happen anywhere."[1]

Braddell's opinions epitomized how ideas about heat had changed since the nineteenth century, from its being an unhealthy element that could impact on human health and wellbeing to something that could be rationalized and managed. Braddell also lived at a time when new technologies such as refrigeration and air-conditioning were entering the market, and this may have changed his perceptions of tropical living. Many historians have explored how tropical heat was conceptualized by Europeans and how these imaginaries informed racialized ideas about the European or the Asian body or the medical study of heat, disease, and the body.[2] Others have written about how heat came to be a foundational paradigm of Western cultural narratives of the tropics, one that epitomized fear and fascination in equal measure.[3] Yet scholars have not yet addressed tropical heat research as it was perceived within the nascent fields of climate and weather science in the Southeast Asian tropics. The few studies that stand out in this area are focused on tropical meteorology broadly conceived and tend to be focused geographically on the Americas.[4]

Bearing this in mind, this chapter explores the science of heat in Malaya and in so doing, perhaps reveals why historians have not focused their sights on this field. It argues that, despite the ubiquity of and the difficulty caused by heat in inhabitants' lived experience and despite the gap in scientific studies of tropical temperature globally, heat studies were not a priority for locally based meteorologists or science writers outside of the medical arena. The one exception was a nascent interest in urban heat as Malayan towns developed over the nineteenth century. This flirtation with the urban heat issue tended to result, however, from the connection of heat and health or was simply viewed as a benign curiosity. There were no major studies of what is today known as the Urban Heat Island (UHI) effect until after the colony gained independence from British rule. Until then it was simply recognized that urbanization increased heat and that this increase might make urban habitation more difficult. That said, the discussion of heat was never far from contemporaries' minds, even if it were not entrenched within scientific agendas.

GLOBAL LIMITS, LOCAL MALAISE

In 1800 heat and temperature research were at a fledgling stage. The concept of geological uniformity across time—as popularized by Charles Lyell on matters of the environment—was still the dominant view of most scientists and influenced thinking about the potential for atmospheric change.[5] As Russell McCormmach has noted, understanding the issue of heat developed only in tandem with the ability to measure it. In 1800, the "fixed point" for calibrating thermometers had not yet been fixed and meteorological thermometry was limited. More than this, the science of measuring air temperature developed significantly only when radiation and the transfer of heat energy were better elucidated in the mid-nineteenth century.[6] Until then, scholars were invested in establishing the thermometric mean of the earth, a discussion that was integrally linked to the notion of atmospheric stasis. Establishing the mean was a preoccupation of François Arago who, in his *Biographies of Distinguished Scientific Men* (which included a substantial section about himself) reported on a scientific debate of some two decades on the subject.[7] Arago described the earth as an "encrusted sun." Unlike our solar system's actual sun, the earth's surface had cooled over time, but it retained within its mass an ancient geological heat and its outer core was warmed by solar rays. His theory was not new but built on scholarship proffered by the Comte du Buffon, Simon LaPlace, and others, who had empirically and mathematically tested ideas on the subject first proposed by Descartes and

Leibnitz.[8] Establishing the mean of the global temperature was a critical part of the discussion.

The discussion of global temperature was nonetheless a Eurocentric one. Fourier's work on the temperature of the whole earth, for example—some of which was published in his mammoth work *The Analytical Theory of Heat* in 1822—conspicuously neglected to mention Southeast Asia, although India and some parts of the Middle East (particularly Egypt) were represented. In this respect his book mirrors contemporary French colonial interests, representing regions and data that Fourier had access to or that were of significance to the government. This is no particular surprise; it mimics the ways in which scientific projects tended to be carried out in imperial countries, focused on areas that would be of benefit to the empire in question.[9]

Contemporary research by Europeans on Southeast Asia was focused on establishing annual temperature averages to determine the mean climate of tropical latitude regions. This was typical of many climate studies in the early nineteenth century, which involved gathering meteorological data to establish patterns, means, and averages so as to deduce the basic laws that were thought to govern the weather, a process facilitated by the establishment of in situ observatories and experimental field stations.[10] For establishing the mean temperature at a certain latitude, one new tool was developed: the meteorological map.

Maps that categorized and depicted climatic conditions around the world have a long history. David Woodward and Brian Harley's work on early cartography as "graphic representations that facilitate a special understanding of things, concepts, conditions, processes, or events" is just as appropriate to the early maps that divided the world into climate "zones" by latitude as to any other form of cartography.[11] Maps based on ideas that dated back to Aristotle or Strabo, whose *Geographica* divided the world into five zones based on levels of heat (with two additional frigid zones) were still influential in the nineteenth century, even inspiring controversial work such as Ellsworth Huntington's maps of "climate energy zones." These were published in his 1915 *Civilisation and Climate*, now lauded as one of the most overt explications of racially driven climatic determinism of the era because of his association of climate with human nature and productivity.[12]

In 1817, however, Alexander von Humboldt represented the work then being conducted on the globe's heat by designing isolines that mapped temperatures onto the earth.[13] These isothermal lines have been described as rendering "obsolete the deterministic anthropologies that Enlightenment philosophers had thought up" (Huntington clearly being an exception to this) and fundamentally changing modern climatology.[14] Isothermal lines were plotted from minimum and maximum (twice daily) temperature observations over the course of one year, taking into account other climatic and environmental influences that might affect those readings such as the type of soil or the proximity of the sea or of mountains.[15] Although ostensibly for a global map, the inclusion of tropical temperatures were still largely the "graphic representations" of Harley and Woodward's depiction, as Humboldt was working with only a few observations from Indonesia to fill in the gap for Southeast Asia.[16] The focus of the maps that grew out of this early work (just like for the work of Arago and Fourier) was the Northern Hemisphere, mainly Europe and America. The maps reveal the paucity of information then known for the tropics, which existed within a broader lacuna of knowledge about the monsoon in Southeast Asia.

During the early nineteenth century, temperature was considered something that could be observed, pinned down, and mapped, but it was rarely considered a changeable or evolving component of large climatic systems. By the mid to late nineteenth century, however, the dominant narratives of climate as being always stable and temperature changes as being temporary and always on local scales were being challenged. Important work here was by scholars such as Louis Agassiz on uncovering the existence of an Ice Age, which threw to the wind the idea of a static climate. Building capacity in this area, the work by mathematician William Thomson (Lord Kelvin) on calculating the cooling trend and eventual demise of our Earth's sun, work by John Tyndall on radiative heat, and Svante Arrhenius's work on atmospheric insulation and carbon dioxide all built around and beyond this.[17]

There were also narratives emerging that focused on the possibility of human-induced climatic change, although these tended to be framed within local and temporally limited scales.[18] For example, the desiccation discussion, popularized by scholars including the Comte de Buffon, also noted how temperature change might be a by-product of deforestation and changed rainfall patterns. Temperature change was not the focus of the debate nor was it considered as important as the possibility of failing rainfall, but it did raise questions. The concept of far-reaching macro- or micro-scale atmospheric temperature changes took time to be accepted into mainstream science, though, and the probabilities and risks were poorly understood at a local level. Julius von Hann—who is considered to have written one of the more influential texts on meteorology for the late nineteenth and early twentieth century—explored the principle of an energy budget deducted through statistical description, connecting climatic fluctuation to historic or cyclical time scales, much as astronomers Annie Maunder and Edward Maunder had theorized before him. But even von Hann, despite his work on fluctuation and variance, gave little credit to the possibility of macro long-term climate change per se, conceiving of it at more manageable scales.[19]

Much of the work that conceptualized climate as a global and changeable phenomenon stemmed from the exploration of extreme weathers, on regional-scale cycles, that had been made possible because of improvements in weather data collection during the latter half of the nineteenth century. Pioneering research was being undertaken, for example, on what would later be termed the El Niño Southern Oscillation (ENSO). In the British imperial context, by the early twentieth century, Henry Blanford of the IMD (established 1874) and successive scientific officers including Rai Bahadur Hem Raj and Gilbert Walker were to develop research on natural variations in the climate system and the relationship of those variations with Indian monsoon rainfall. In Australia, Edwin Quayle was also working on climatic oscillation (later known as the Southern Oscillation Index) and long-range forecasting at the Melbourne Observatory.[20]

The last decades of the nineteenth century also saw the widespread usage of synoptic mapping, which had developed after the 1850s, reliant on the advent and institution of telegraphy.[21] These differed from the meteorological maps of Humboldt's era as they incorporated "real time" information derived from recent observations, which were fundamental to predicting weather. The physical practice of plotting simultaneous observations enabled teleconnections to be visibly revealed at local and regional scales but also aided in the other significant concern of the yet undeveloped field of tropical meteorology: typhoon prediction.

Typhoon prediction was not then considered to have an obvious connection to heat. Early work on typhoons was directed to establishing laws and principles on their formation and motion. The normal focus was on pressure as an indicator of an impending storm.[22] The synoptic map was a later significant tool in predicting the path of a typhoon; it was much used in the observatories that bordered the China Sea, such as the Jesuit observatories in Manila and Shanghai and the British observatory in Hong Kong.[23] The advent of synoptic maps did, however, enable a visual and dynamic plotting of temperatures, including heat waves, which, it was realized in much later research, were indicators or symptoms of incoming typhoons in the tropics.[24]

Temperature studies also suffered from a disjunction between atmospheric studies and the field of meteorology. Prior to the Second World War, the nascent field of practical meteorology was divided into three camps: forecasters, who aimed to improve weather predication accuracy; theoretical meteorologists, who grounded deductive knowledge in mathematical law and physics; and empiricists, who induced weather laws from data analysis rather than from the mechanics of physical qualities.[25] Much of the discipline remained focused on data collection, statistical analysis, and forecasting, which meant that exploring global climate phenomena or even regional and local climatic changes was not a priority for meteorologists. The inability to make widespread upper-air observations until the aviation industry took off also meant that much atmospheric temperature research was literally grounded. Temperature studies of, or in, the terrestrial tropics were therefore more often derivative of other interests, rather than the focus of research for its own sake.[26]

TAKING THE COLONY'S TEMPERATURE

That said, there were some early investigations into temperature in the Straits Settlements, with observations made in Malacca in 1809, 1815, and 1816; in Penang in 1786–1787 and 1822; and in Singapore from 1820 to 1824 and a few in the 1830s and 1860s alongside rainfall.[27] That these short observational runs were not continued is perplexing, but there may have been cogent practical reasons for not continuing them that have not yet been mentioned. Temperature instrumentation was delicate, complex, expensive, and easily broken. It is likely that some of the early abovementioned observations were made using instruments on loan from EEIC vessels then docked in port, rather than on separate sets of instruments made for the new settlement as part of a terrestrial observation project. Given the specialist nature of this equipment, if breakages happened, the expertise and parts were not available locally, so all new parts or instruments had to be ordered and shipped from London, a process that could take years from order, to creation, to eventual shipment.[28]

In the 1850s, the process was refined, with Kew Observatory becoming the center for the verification and standardization of thermometers to the Fahrenheit degree scale. These thermometers were then distributed to EEIC ships, British imperial observatories, and colonies in a more coordinated effort to provide the empire with the scientific instrumentation to chart its atmospheres, although all instruments were still manufactured and shipped from London in a long and complex logistical process.[29] The story of the Singapore observatory's thermometer is a case in point. Under Charles Elliot, all observatory instruments were specially commissioned to order. They were made in London and from there shipped to Aden, to Bombay, and to Madras where they were calibrated. Only then were they transported to Singapore on the EEIC ship *Minerva*, the last part of the journey in Madras and to Singapore alone lasting three months.[30] This is in direct contrast to rain gauges, which were relatively simple instruments that could be packed and transported easily or even manufactured locally. These practicalities may explain the inconsistencies and gaps in temperature observations in the Straits Settlements before the 1860s.

The small sets of pre-1860 observations were put to a much bigger purpose, however. David Brewster, the mathematician and physicist then responsible for the editorship of the *Edinburgh Journal of Science*, scrutinized the Malayan observations in his quest to determine the mean temperature of the equator. Brewster was especially interested in the Straits Settlements' temperature readings as they represented the closest geographic location to the equator then obtainable (only 137 kilometers), as well as comprising multiyear datasets, which were very unusual at that time. Before this, Brewster had only meteorological readings made in Batavia and Ceylon to work with for Asia.[31] The Straits Settlements' readings helped him to close a gap. Determining the mean temperature of the equatorial region specifically was a question that Alexander von Humboldt was considering in the 1820s, in connection with the wider discussion of the global mean.[32] This built on his work with isolines, which took into account the problematic interference of mountain ranges and valleys that impacted temperature at different heights. However, something of a dispute had ensued between Humboldt and the astronomer and mathematician Henry Atkinson, after they arrived at different conclusions for the estimated mean temperature of the equator. Atkinson accused Humboldt of making an error, and vice versa. Thus, the importance of the observations from the Malayan Peninsula becomes apparent, as they enabled Brewster to work with geographically more appropriate datasets than his contemporaries and to update and resolve the argument. The new information also absolved both Humboldt and Atkinson from error, as they had not worked with authentic regional observations, only estimates.[33]

This whole episode, and the weather observations, was played out publicly in a range of scholarly journals, including the *Edinburgh Journal of Science* but also in newspapers local to the Straits, such as the *Singapore Chronicle*. Their dissemination across such a diverse array of scientific and popular mediums positioned the Straits Settlements at the heart of a global scientific debate led by geographically diverse and scientifically astute characters. Regardless, although the discussion may well have elicited interest in the colony, local actors did not contribute publicly to this debate. This is perhaps a surprise, given the proclivity of nineteenth-century Renaissance men to contribute to scientific or pseudoscientific journals. Even William Farquhar, for example, to whom the early Singapore observations were credited, did not have a say; nor did his contemporary EEIC Bengal Native Infantry officer Lieutenant Charles Davies, who likely made the observations on Farquhar's behalf. Farquhar is better known for commissioning extensive collections of natural history drawings.[34]

Neither did contemporaries make much use of these observations, at least not until much later. James Logan's later conjecture that the large-scale destruction of Penang Island's virgin forests could affect temperature as well as rainfall was not based on local temperature readings, for example, but on extrapolation from cases elsewhere in the world, such as the Cape de Verde Islands (modern Republic of Cabo Verde), which now suffered a "sultry atmosphere and dreadful droughts" owing to deforestation.[35] Aside from the lack of instrumentation available for undertaking long-term sets of weather information in these first few decades of the colony's existence, the lack of trained scientists and meteorologists may well have hampered further discussion.

Gentlemen practitioners were the norm in Europe as well as in Malaya, but there were no local equivalents of a Humboldt or a Brewster to promote research into the climate for its own sake. The most celebrated minds in Malayan physical exploration and scientific writing of this era were focused on subjects including botany, natural history, and geology. The prolific superintendent of the Calcutta Royal Botanic Gardens Nathaniel Wallich, for example, spent his time in Singapore devoted to collecting plant specimens, as did the biologist, explorer, and naturalist Alfred Russel Wallace, although Wallace did think geographically and climatically around the question of the distribution of animals globally.[36] More importantly, although both men are associated with science in Malaya, neither one stayed in the colony for long. They were only visitors to the peninsula.

The discourse concerning man-made climate change and climatic instability that emerged during the nineteenth century was slower to take hold in the tropics, and (unlike rainfall, which was gradually accepted as a potentially erratic and unpredictable phenomena at a local scale) heat continued to be considered a relatively stable feature of the tropical climate across the year. Captain James Low of the Madras Engineers had argued in the 1830s that there was little temperature variation across the peninsula, perhaps one degree difference between the east and the west coast. If people experienced different levels of thermal comfort, this was because of proximity to the ocean and to sea breezes.[37] The medical doctor Robert Little had also decided that "there is a great uniformity of temperature throughout the entire year" and Staff Surgeon John Buckley, explained how "the degree of temperature is constantly very high."[38]

Even by the early twentieth century, the idea of year-round stability was still current, albeit more within popular culture than in scientific literature. For instance, Charles Burton Buckley, newspaper editor, antiquarian, and long-term resident of Singapore, wrote in 1902 that "a Sumatra squall blowing away the *attap* roof of a house on a hill overlooking the present Ladies Lawn Tennis Ground . . . some thirty-five years ago, on the morning of the day when there was to be a dinner party in the evening as a housewarming, is, possibly, the worst that can be alleged against the forces of nature."[39] The gradual erosion of this idea in the scientific literature does become more apparent in the first decades of the twentieth century, however.

The November 1903 edition of the *Agricultural Bulletin of the Straits and Federated Malay States* reveals the juxtaposition between popular culture and scientific discourse. The journal had printed a letter concerning the possibility of growing cotton in Malaya, based on the US experience. The letter writer, one A. S. Baxendale, an overseas observer, wrote to the federal secretary at Kuala Lumpur noting how the extreme weathers that caused huge losses to the cotton growers in America would never cause serious losses to Malayan planters.[40] A retort, printed a month later in the *Straits Times* and written by Stanley Arden, superintendent of experimental plantations for the FMS whose views might be considered more informed, argued that "assuming the climatical conditions here could be considered suitable, which personally I very much doubt, remembering that the monsoons are scarcely discernible and that the driest month of one year may be the wettest of the next."[41] By the twentieth century then, although ideas concerning climatic stability had shifted, many ideas persisted in the popular imagination about Malayan climate, and any sense of the changeability and unpredictability of the climate was dominated by the discussion of rainfall, not of heat. This final point was to subtly change in the early twentieth century, as the experience of living in the rapidly growing urban areas thrust a new set of challenges in the way of inhabitants and scientists.

"CIVILIZING" THE JUNGLE AND CREATING AN URBAN ATMOSPHERE

In the late 1700s, James Six—a man widely recognized for his pioneering work in inventing an effective thermometer that registered minimum and maximum temperatures—had undertaken a series of experiments on urban temperature that were read to the Royal Society.[42] Six had become interested in variations in urban temperature in his English hometown of Canterbury.[43] Although he did not investigate how towns and cities had different temperatures from rural areas, he did note how urban form affected air flow in unnatural ways, thus impacting on temperature at different locations, heights, and times of the day. These experiments foregrounded those such as Luke Howard undertook in and around London in the 1810s.

Howard was a trained pharmacist and self-proclaimed meteorologist who became an influential figure in Victorian weather science. He was perhaps better known for his work in developing a standard nomenclature for cloud types, but he also undertook many studies on London's climate. Much of this work entailed extensive instrumental observations of the weather in and around London between 1806 and 1830, with results from which he read in a series of papers and published in the three-volume work titled *The Climate of London*.[44] This seminal text incorporated his analysis of temperature readings made at multiple locations within and outside London, which quite clearly revealed differences in temperature between London city and the adjacent rural villages. He concluded that "London is always warmer than the country." He attributed this phenomenon to a twofold "artificial" interference on temperatures. First, human activities such as domestic fires or burning coal to power industries created heat, but densely packed buildings and narrow streets worked to inhibit natural air flow that prevented heat from being blown out and away from the city.[45] He also identified several factors that are now commonly recognized in modern day urban climatology, such as the concept of an urban canopy and that of a street canyon, which form a distinct close-scale urban climate.[46] Similar theories were also developed and explored by Emilien Renou in Paris during the 1850s and later by Victor Kremser in Berlin in the 1880s.[47]

Many of their ideas crept into the Straits Settlements through the movement of people, knowledge, and text. We see their influence in the work of colonial officials who worked across the British Empire and came to spend time on the peninsula. John Crawfurd, for example, a medical doctor, amateur scientist, and onetime British Resident at Singapore, was another Renaissance man who looked at climate amid his other interests. In connection with his medical studies, he dedicated an early part of his research to understanding the temperature of the island. He came up with an interesting conclusion. Based on comparison of annual averages from across the British Empire, he realized that the temperature in Malaya was far lower than in other areas with the same tropical latitude. Having landed on this anomalous finding, he continued investigating. Undertaking this work in the 1850s, he had several decades of scattered weather observations to work with and discovered that over twenty years, the temperature, though lower than elsewhere in the tropics, was steadily increasing. The influence of Howard's research on Crawfurd's work was clear when he explained this increase as being "ascribable no doubt, to the increase of buildings, and to the country having been cleared of forests for several miles inland from town, the site of the observations."[48] He went on to note how the lower maximum (in the shade) of the older observations made by Mr. Thomson (82.3°F as compared to 86.7°F at Crawfurd's time of writing) would have been affected by the nearness of the forest to the town, which would have caused a "greater degree of coolness." Yet he was also concerned as to why the annual minimum then was in fact higher than at the time he was writing. He postulated that the forests may have had a dual effect, while bringing down the maximum, they had also "hindered the free circulation of air and consequent evaporation."[49] His conclusions were simplistic, but they reveal the application of ideas about urbanization, the movement of air, and the loss of natural vegetation on thinking about heat.

His views were not at all unusual. John Cameron, owner and editor of the *Straits Times* from 1861 observed that, "by resorting to the neighbourhood of the jungle, a degree at least of reduction in the temperature may be secured. In such places as Selita [sic] . . . and with the primeval forest lying all around them, the additional coolness is palpable, and cannot be less than two or 3 degrees."[50] When Alexander Maingay published his strong views on deforestation and desiccation, he also mentioned that "the ruin of the forest . . . involves . . . a general elevation of the temperature," and PCMO Allan Maclean Skinner also explicitly linked a perceived rise in temperature and the increasing prevalence of droughts to "the spread of buildings" as well as to the loss of forest cover.[51]

Skinner had access to what were by that time fourteen years of continuous weather records from the Straits Settlements, collected under his own watch at the Medical Department, and he compared these with tables of weather observations that had been furnished by the director of the Batavia observatory for 166 registering stations in the Eastern Archipelago. He was interested in the possibility of a connection between human-induced climatic change and heat and in the relevancy of sunspot activity for temperature fluctuation. He was convinced of the relationship between deforestation, urbanization, and temperature increases, and he discussed a theory then in common currency that sunspot activity had an approximate eleven-year cycle, which correlated with changes in rainfall, storms, and temperatures across the globe. Drawing from the work of men including Joseph Baxendell, Piazzi Smyth, George Airy and Wladimir Köppen, Skinner postulated that, in tropical latitudes, heat waves generally fell the year before maximum sunspot activity and that the "regularity and magnitude" of this temperature "wave" were very strongly marked in the region.[52]

Aside from the journals that published on aspects of local climate, flora, fauna, geology, and geography such as Logan's *Journal of the Indian Archipelago and Eastern Asia*, a group of colonial officials founded a *Straits Branch of the Journal of the Royal Asiatic Society* in 1878 modeled on the journal created by Logan. The journal's membership and beginnings were closely tied to the Raffles' Museum and Library institution and staff.[53] Many of the prominent men in colonial Malayan society with connections to meteorology were frequent contributors and members, such as W. E. Maxwell, T. Irvine Rowell and A. M. Skinner. This journal was a key outlet for publishing literature on aspects of meteorology, meteorological observations, and musings on climate changes and their causes. The journal was accessible to the reading public through the Raffles Museum and Library (established in 1874) alongside their regular collections, which were heavy on educational and nonfiction subjects and included many books relating to the sciences and natural history.[54]

The reading public also had access to scientific or semi-scientific opinion pieces and articles that were frequently reprinted in shorter form in the newspapers, including several dailies and weeklies such as the *Singapore Free Press and Mercantile Advertiser* (from 1835) or the *Straits Times* (from 1845). It is perhaps no surprise that there was a colonial elite whose members and affiliations overlapped in the ownership or editorship of the newspapers and journals and also in their membership of organizations such as the Straits Branch of the Royal Asiatic Society, the Raffles Library, or the horticultural societies. These elite were often the same as the gentlemen observers making weather observations in the Straits Settlements or contributing articles on meteorology to the journals.[55] This meant it was likely that a proportion of this colonial class were exposed to Western scientific ideas about temperature and urban heat—for example, the British lecture reprinted in 1907 in the *Eastern Daily Mail and Straits Morning Advertiser* on the "effects of civilisation on climate." This echoed Howard's claims from almost a century earlier of how buildings raised the temperature of the urban locality.[56] Moving on to wonder whether planting more trees in Singapore was a desirable solution, a commentator on the 1907 piece continued, "it is impossible to deny that man and his works do influence climate to a greater or less extent; the spread of civilisation in a new land has a real effect on the annual tale of weather."[57]

HEAT WAVES IN THE URBAN JUNGLE

By now, the idea of a specific urban heat was common currency. So, while scientific temperature studies had taken a back seat in Malaya, the experience of heat that was dominating many inhabitants' experiences of living in tropical towns was changing how people thought about tropical temperature, especially as the towns grew. In March 1905, for instance, town temperatures had hit a record 93.5°F. "The weather is much hotter now than it has been for a very long time," the *Straits Times* reported.[58] Another hot spell in 1914 led to a relaxation on rules for school uniforms, with boys permitted to remove their coats during school hours in the absence of any fans being provided at Singapore's schools.[59] Extreme heat was remarked upon for its influence on productivity and health, but the dry weather that normally accompanied heat lead to other problems as well. In 1907, for example, a fire broke out in brushwood on waste ground "between Bidadari cemetery, on the Serangoon Road, and the junction of McPherson Road" in Singapore:

> The recent hot weather had made the jungle and undergrowth very dry so that the flames spread with great rapidity, and a large area was soon in flames. A row of shop-houses on the Serangoon Road and several native huts were in danger. The police were communicated with, and Inspector Crummey and his assistants attached a couple of lengths of hose to a hydrant and drenched the *attap* roof of these buildings. As the fire continued to spread, the Brigade

> was summoned to Bukit Timah Road, and Superintendent Petl with the motor engine Brodrick, and accompanied by a number of men on the tender, proceeded to Bukit Timah. There was no fire there and after making inquiries by telephone, they ascertained that the outbreak was at Serangoon Road. They then assisted the police in saving the houses.[60]

The colonial towns of the Malay peninsula had never been free of fires, especially in the early nineteenth century when a lack of building regulation and a tendency to build in wood and palm created highly combustible environments. Both George Town and Singapore had lost and rebuilt large sections of their early settlements—including Singapore's Kampong Glam, which had burned to the ground in 1847, or the Tanjong Pagar docks, which had been destroyed after a stray spark ignited coal sheds during the drought of 1877. As building materials became more durable and fireproof, however, and regulations improved, so the incidence of urban fires decreased.[61] But from the 1920s onward, heat waves were reported with increasing frequency, and there is a rise in fires that contemporaries attributed directly to these periods of extreme heat and the associated dry conditions.

The term "heat wave" is a relatively modern one. Its first known usage was in the 1840s, possibly derived from the work of the polymath John Hershel.[62] It gradually entered contemporary vernacular and is first seen in the Malayan newspapers in 1899 but appears infrequently until the 1920s when it was commonly adopted in reference to unusual hot spells.[63] Placing this into global context, it was during the 1920s that the subject of urban heat also became a dedicated scientific study. Publications emanated from the pens of researchers on the towns of the American tropics in the 1920s. In Asia, Japan led the way with major studies on heat in Tokyo and Osaka in the 1930s.[64] In Germany, the 1937 publication of Albert Kratzer's *Das Stadtklima* proved a pivotal moment for European urban heat studies and led directly to some of the terms and theories still used today, including the conceptualization of a distinct urban climate, or "meso-climate," comprised of often vastly different micro-climatic zones. Zones might span industrial or residential areas or, on a smaller scale still, streets, parks, and squares.[65] The models determined in the 1930s underpin our modern fundamental frameworks for a UHI: local scale (similar buildings in a district), microscale (urban canopy), and mesoscale (total urban boundary layer).[66]

During this period Malayan towns underwent many changes that amplified the propensity for urban heat. Urban populations rose because of immigration, especially during the interwar years, with the largest proportion of these migrants from low-income backgrounds seeking work opportunities in the colony. Many new arrivals lived in the most densely packed areas of the towns in multiple occupancy housing.[67] Successive waves of economic depression across the 1920s and 1930s left many of these new inhabitants very vulnerable, and their low income and overcrowded conditions meant they were at risk from sudden extremes of heat.[68]

The urban landscape was in a slow process of metamorphosis necessitated because of the motor vehicle. Traditional streets were built to withstand foot traffic, rickshaws, the occasional horse, animals being driven to markets or to slaughterhouses, and later, rubber-wheeled *jinrickisha* pulled by "coolies," the contemporary term for manual laborers. These streets were generally made from packed earth or laterite. They were "watered" by an army of prisoners or low-cost "coolie" labor during dry spells to keep dust levels at a minimum. Between 1915 and 1925 though, the number of motorcars in Singapore alone had swelled from 842 to 4,456.[69] These vehicles were almost useless when packed earth streets turned to mud during heavy rains, and so gradually the roads had to be improved. Surfaces needed to be made more durable, and streets had to be widened to facilitate traffic. One road surface material that was experimented with was natural rubber latex, capitalizing on the booming local industry, but mostly roads were "metaled" using a combination of granite, limestone, and bitumen (asphalt).[70] By 1920, 96.26 miles of metaled roads had been fabricated on Singapore island, and the municipal councils of George Town and Singapore steadily financed more mileage over the 1920s and 1930s.[71]

In modern studies of the UHI effect, both granite and asphalt have been found to retain and increase city heat. Asphalt has high surface temperatures and granite, while feeling cool during the day, retains its heat overnight, the time when UHI signatures are most apparent.[72] Both materials are significant contributors to the UHI effect, as are the cars driving along them. Although the impact of car usage and roadways ought not be overemphasized, Malayan towns were still small and relatively rural as compared to today, there would have been without doubt a change in the urban experience of heat. Widening roads frequently occurred at the expense of tree-lined verges, grassy roadsides, and even parks. Reminiscing about how Raffles Place (the past and present heart of Singapore's central business district) had looked in 1901 from a standpoint of the 1930s, for example, local inhabitant Edwin Brown recalled how the commercial square had been transformed—from an attractive green spot shaded by "flame-of-the-forest trees," where men would sit to eat their lunch and horses and carriages would shelter "while the 'mems' did their shopping" into a "glaring whiteness . . . with its motor cars parked in orderly ranks in the centre."[73] His statement is clearly borne out by contemporary photographs of the area, which chart the steady shift from an attractive green space to a treeless and practical parking lot.[74]

Brown's description clearly accompanies the rise in press reports of heat waves and excessive nighttime temperatures. In May 1935, Singapore's inhabitants experienced what was "probably the warmest day . . . this year and many were the grumblings and grousings heard as large handkerchiefs were laboriously passed over sweating brows and clammy necks. The day was swelteringly hot and the fact is borne out by the figures of the Meteorological Office . . . the maximum temperature registered was 90 degrees in the shade."[75] In 1938, another heat wave was reported, with inhabitants of Singapore unable to sleep because of exceptional overnight temperatures of 80°F.[76] Reports of hot-weather-induced fires, rather than ones simply ignited through mishap, were commonplace. In February 1935, for instance, there were frequent "grass fires . . . on account of the hot weather," the *Malaya Tribune* reported. "There have been several cases during the past three weeks and a further three fires broke out in different parts of Singapore yesterday, most of which were on areas of waste ground."[77]

February is normally hot in Singapore, but in 1935 the hot conditions persisted across the year with grass fires breaking out across the town during August, "owing to the terrific hot weather the colony is now experiencing. In all cases the outbreaks were due to spontaneous combustion."[78] Spontaneous combustion continues to be reported after the Second World War ended, bringing more hardship to the town during a difficult phase of postwar recovery. In July 1947, for example, more than an acre of *lalang* (grass) fire threatened local storehouses and, in the absence of fire engines, had to be put out with buckets of water.[79] Then, just one day later, the hot weather "caused two more lallang fires in the outskirts of Singapore [when] four to five acres of lallang, beside the Admiralty Oil Farm, at the 4½th Milestone, Ayer Rajah Road, were burnt out in a fire which broke out at 4.30pm. In the other fire, lalang [sic] at the RAF Aerodrome, Changi, was set alight," necessitating the attendance of the Singapore Fire Brigade.[80]

In April 1948 temperatures were excessive. "Phew . . . it's hot!" proclaimed the *Morning Tribune*, telling how "sweat-drenched town dwellers pulled their swimming costumes out of their trunks and made a beeline from the town areas to the beaches in Katong, Changi, Ponggol, and other points."[81] This heat wave, falling across the Chinese Qingming festival, led to a record outbreak of fires: twenty-two over four days. Many of these fires broke out at cemeteries where unattended joss sticks, left during traditional celebrations for the ancestors, set light to dry grasses and weeds. "The heaviest day for the Brigade was Sunday, when there were 10 fires, eight of which were grass fires at or near Chinese cemeteries," noted the *Straits Times*. "There were four grass fires on Saturday, one of which at Bukit Brown Cemetery, took four hours to subdue."[82]

These reports stand out because the temperatures were considered above and beyond expected norms for the seasons of the year, even for the normal intra-monsoonal periods that are customarily hotter, rather than simply being complaints about the tropical heat per se. The temperatures during the 1935 heat wave, for example, were above the average for the time of year, qualifying as a heat wave of about three weeks and were accompanied by extraordinary nighttime highs.[83] In 1936 too, "one of the severest heat spells ever recorded in Singapore [sent] long standing residents of the colony [to] describe recent weather conditions as the most trying within their experience. On 21 nights in the last five weeks the minimum temperatures recorded in Singapore exceeded 76 degrees. On two nights the minimum was 80 degrees." Even C. D. Stewart, superintendent of the MMS and a man not prone to hyperbole, told the *Singapore Free Press and Mercantile Advertiser*, "I do not remember another month like it." Stewart noted:

> The combination of high temperatures and still conditions has made the weather exceptionally trying. High night temperatures not only make it difficult to sleep. They impair the restfulness of sleep and make it more difficult for the body to withstand the heat of the day. In this connection it is noteworthy that a very slight variation in temperature may mark the difference between a relatively cool or hot night. . . . In Singapore the average minimum night temperature is somewhere about 75 degrees. Anything above this becomes perceptibly unpleasant, and is accompanied by considerably more ill-effects than any similar variation in the maximum day temperatures.[84]

The following year, the *Malaya Tribune* raised the subject of everyone's current obsession with the weather. "Everybody talks about the weather in Malaya," the writer posited. "We wonder whether it is hotter this year than last, how long the drought will endure, and so on."[85]

Although it is tempting to ascribe the uptick of reports to early climatic changes and, certainly, the narrative evidence is compelling, the changes that contemporary inhabitants were picking up on may not have been inspired by a local or regional temperature rise. Indeed, while C. D. Stewart complained about the extreme heat in 1936, he produced a report that same year to suggest that the heat wave was an anomaly and that temperatures were relatively consistent over the years. Rarely did air temperatures hit the excesses of other parts of the tropics. With the benefit of greater hindsight than was available to Stewart, a longer-term look at annual temperature averages supports his conclusions, with the period from the 1820s to 1941 holding an annual average of around 26–27°C for night and day combined.[86] Data collated after the Second World War by the modern Meteorological Service Singapore (the post-independence successor to the Malayan Meteorological Service), part of Singapore's National Environmental Agency, suggests that this pattern held until the 1970s, when temperatures began to accelerate, driven up by far more extensive and very rapid urbanization.[87]

Explaining the consistent upward trend in the historical narrative since the 1920s and 1930s of reports of heat waves, rather than simply complaints about the tropical heat and humidity in general, especially those that discuss nighttime temperatures, is thus complex and may have been driven more by changing thermal discomfort in the towns than by changing temperatures per se. The density of buildings and problems of air flow were raised time and again in early twentieth-century planning and public health discussions, with initiatives to improve housing, widen back lanes and alleys, and increase the provision of public parks all on the agenda.

In the 1920s, the editor of the *Straits Times* was asked his opinion as to Singapore's urban future. In a particular resonant statement, he opined:

> The whole island of Singapore is destined to become urban, and because its geographical location is tropical, the planning of its streets, the provision of its open spaces, the design, arrangement, and distribution of its houses, shops, *godowns* and factories are matters more vitally important than they would be in a temperate zone. Great main arteries, broad and handsome, should be formed, not mere patchings up of existing roads, but lined so that the cleansing and refreshing breezes that rise on the sea may sweep along them unrestrained, sweetening and purifying as they go. Crossroads should be wide also and dotted here and there at frequent intervals; wherever the ground is favourable, there should be public gardens, with trees and shrubs and flowers to give restful shade and to kindle in the minds of the people an appreciation of Nature's beauties. . . . in the long run it will be the . . . surest of all ways to promote health, prosperity, and content.[88]

With clear echoes here of the turn-of-the-century "garden-city" ideal, in this statement he reveals contemporary ideas on the responsibilities of urban planners not only for improving working inhabitants' lives through bringing nature into the city but also for improving the experience of heat in the tropical city. Although the local government in Malaya did not go to such lengths to import garden city ideals into Malayan towns before the Second World War as they did elsewhere in the colonies (Hong Kong, for example, was the site of a new model suburb based on Ebenezer Howard's garden city principle), the investigations of men such as Edwin Chadwick, Charles Booth, and Seebohm Rowntree made a significant contribution to changing public opinion toward the role and responsibilities of government in improving the quality of urban life in Britain and, later, in the colonies.[89] The advent of the Singapore Improvement Trust in 1927 was a significant step in this direction with its charge of improving urban infrastructure, opening or constructing back lanes, improving living conditions, and delivering new public housing schemes.[90]

Nonetheless, the issue of thermal discomfort did not go away, so it may be considered strange that the first comprehensive UHI study of Singapore would not be published until almost a decade after the nation achieved self-rule in 1957. This study, undertaken by S. Nieuwolt and a team of students, involved a series of dry- and wet-bulb temperature measurements at different locations in and around Singapore to assess the extent of the effect of urbanization on temperature. This was probably the first, and last, experiment to be undertaken on the UHI effect until the 1980s.[91] This neglect under the colonial government may have been due to the preoccupations of the prewar and postwar periods; the small scale of urban development and relative low-rise infrastructure of Singapore relative to cities where manifest studies had been undertaken (say Tokyo or Berlin), or simply the preoccupations of the meteorological service, whose small team of officials were focused on providing weather reports and forecasts for military operations from the Malayan Emergency of 1948 to 1960. It may also have much to do with the fact that the steepest period of urbanization in Singapore and in the towns and cities of Malaysia occurred only after the period covered here, after the two countries became independent from one another in 1965.

In Singapore particularly, the 1960s saw a swift and forceful move toward modernization, widely credited to the vision of Prime Minister Lee Kuan Yew and the ruling People's Action Party. This came in the form of sweeping economic reform, rapid urban infrastructural development, and industrialization, all of which had a rapid impact on urban temperatures and pollution.[92] Now combined with irrefutably rising local, regional, and global temperatures, the issue of thermal comfort has subsequently become one of the most pressing issues of the twenty-first century cities in Singapore and Malaysia, as in every city in the world.[93]

PART III

CONTROLLING WEATHER

This final section of the book explores tropical weather extremes and the means to control them. It draws extensively from recent work undertaken on controlling rainfall (and thus engineering a colonial environment) by scholars such as Philipp Lehmann and Meredith McKittrick and, as an extension into the micro-world of climate and environmental control, the work on indoor atmospheres by Yuriko Furuhata.[1] There is also a wider literature on weather modification that bears influence here, much of which is based on new scientific thinking that emerged in the late nineteenth century (and more from the 1920s to the 1950s) that advocated deliberately modifying weather to improve local outcomes such as agricultural productivity.[2] It is an interesting point that, despite the presence of the British on the Malayan peninsula, who had themselves been part of many contemporary enviro-technical debates and, indeed, were practicing some of their own, including rainmaking activities, in Malaya these debates were not advanced.

In many respects this was a reflection of the state of Malayan science, which cannot be separated from the limited funding that the British government wanted to invest locally. "The characteristic features of the climate of Malaya," wrote S. G. G. Kelliher in 1936, "are uniform temperature, high humidity and copious rainfall."[3] Thus, the direction of science in Malaya was aimed solely at managing the challenges presented by these conditions. The primary desire was to manage tropical extremes, and this manifested in engineered projects to manipulate water or to remove heat and humidity—despite evidence in the press of scientific discourse around reforestation as a means to alter the weather or awareness of new tools for weather modification.[4]

Flooding was one of the earliest issues to be tackled, largely because of its pressing urgency. The new towns of the Straits Settlements were developed on low-lying and often swampy land in coastal areas that were prone to tidal and riverine flooding, sometimes multiple times a year. Flooding was also a problem in rural areas, but this seems to have entered the colonial narrative less, unless it directly involved the destruction of a plantation or destroyed a rice harvest.[5] The irony is that the colonial economy encouraged large land clearances for plantations, which in turn increased the propensity for flooding or exacerbated the erosion of exposed earth during droughts.[6] Droughts were unexpected because of what was assumed to be a high all-year-round rainfall, but actually, droughts are common in the tropics, a fact that became apparent through lived experience and better keeping of long-term rainfall records, through which patterns in their occurrence came to be identified.

In the nineteenth century, the management of floods, droughts and heat tended to be low-tech and ad hoc. By the twentieth century, the professionalization and departmentalization of the colonial service—combined with the advent of new experimental and research stations for forestry, agriculture, soils, and a countrywide Drainage and Irrigation Department—enabled better solutions for managing the effects of extreme weather on the environment. In the towns, major engineering projects to store and pipe water were underway, and in the civic, medical, domestic, and entertainments spaces of the evolving settlements electricity, fans, and air-conditioning were revolutionizing tropical living, for some.

CHAPTER 5

PLANNING FOR RAINY DAYS

Our built environment exists as a partial reflection and physical reply to the extremes of the local atmosphere and environment. Buildings tether the natural landscape, inhibiting natural flows of water, blocking out light, heat, and cold. Infrastructure provides the arterial network, existing as liminal spaces and systems that cross the constructed and the organic—running under, between, or over places, and blurring the boundaries of natural and human worlds. Paul Edwards argues for the importance of such infrastructures in validating our climate knowledge; a viewpoint especially relevant to this chapter, which explores how the infrastructural design of colonial towns developed in response to the excesses of water.[1] The argument centers on a chronological investigation of flood management in the urban and peri-urban areas of the Straits Settlements, considering how and why top-down responses to rainfall extremes changed over time and how water shaped social vulnerabilities and resilience in response. In so doing, it sheds light on how colonial authorities understood the problem of flooding in terms of needs-based responses rather than as a point of meteorological investigation.

Rohan D'Souza has argued that modern historiography of imperial water management has conceptualized rivers and floodwaters as geomorphological processes that butted against the colonial technocratic vision and intention to "suppress, if not eliminate," seasonal or natural inundations within "dynamic hydraulic environs."[2] His statement reflects Reyner Banham's contention that the "deployment of technical resources . . . control the immediate environment."[3] Such discourses have been tempered in recent years, but it is true colonial government officials did make many attempts to rein in the immediate environment and to control the weather around them, in the name of economic protectionism, public health and safety, or simply just in the quest to create more functional urban environments.[4] This chapter argues, in tune with previous chapters that explore colonial responses to weather, that water control was largely a result of the day-to-day practicalities of creating a livable but (importantly) an economically viable environment that would not be subject to repeated disaster. This was less a result of coherent forward planning, however, than a series of responses to challenges as and when they arose. In this reading, the colonial reach over the environment was ad hoc, reactive, and occasionally ineffectual and reveals a lack of colonial technocratic ambition. Environmental colonialism was, put simply, a leaky vessel.

Contemporary knowledges of tropical atmospheres hid in the background of this infrastructural urbanscape but were not always part of the dialogue. There were multiple disjuncts between developing climate knowledge and its embodiment in the built environment that ought not to have existed, as the course of water management evolved out of a mix of customary engineered solutions (in this case often imported from Europe and adapted) and a healthy dose of practicality. The praxis and embodiment of climate knowledge within engineered and technological water management have often been considered a reflection of imperial perceptions of foreign (especially tropical) weather and environment, and in this instance, the lack of interest in local climatology may well be a case in point.[5]

Relatedly, this chapter also considers how flooding and British colonial water management also affected the poorest in society and impacted indigenous practices of water management. This was the result of the deeply embedded socioeconomic structures that governed access to clean water (usually predicated on money and the perceived value of particular social groups) and the ideological preference for adopting Western management techniques as the modus operandi. The ideological frameworks that governed these processes are termed "colonial" in this context but are also representative of broader inequalities and value judgments in societies at large.

Bringing these themes together, the chapter addresses the notion of environmental control from the standpoint of water, rainfall and its extremes. It explores how the colonial government integrated climate knowledge into their water management strategies and the extent to which such strategies were simply responsive, rather than proactive or part of a grand design to govern the local environment. In so doing, it raises the spectre of the unpremeditated consequences on colonial society which, intended or no, served to reinforce the inequalities extant under British rule.

PLANNING A TOWN IN THE TROPICS AND THE PROBLEM OF TOO MUCH WATER

In the nineteenth- and early twentieth-century towns of the Straits Settlements, the most frequent time for flooding was (and it still is) during the northeast monsoon with its frequent and intense heavy rains peaking between December and January. The low-lying and often swampy coastal and riverine regions of Singapore and George Town experienced tidal and riverine flooding especially when the rains coincided with a high tide. Urban flooding events were not only meteorological phenomena, however, but an inevitable consequence of too rapid and poorly constructed or planned urban development. The scale of an event—whether it resulted in an inconvenience or a disaster—was connected to the preparedness of the town at any given time, whether that be through ample flood defenses or social resilience, usually a combination of both. Disasters did not—and do not—happen in fully prepared or unpopulated areas.[6]

The contemporary situation was neatly described by James Scott, a merchant trader who lived and carved out a fortune in the early days of George Town. "Our present station," he wrote in 1794, "is in a sand Bank nearly Insulated by a Creek. The Bank itself is intersected with salt Nullahs, and is in most places 12 inches below Equinoctial Spring tides, and where highest about 18 inches above them. Hence it presents in the rainy season, at high water, a curious Scene of Houses and Streets and hillocks intersected by or surrounded with water."[7] Early descriptions of Singapore are similar, with Dr. Robert Little noting in 1851 how the settlement was riddled with water courses, from three main rivers to an array of streams and rivulets that drain toward the sea. The Rochor River, which was altered early on to form a canal, "carries off the superabundance of water that would otherwise stagnate on a valley of some size through which the Bukit Timá [*sic*] road runs . . . at Campong Malacca, the river divides into numerous branches, one of which proceeds as far as the Sepoy Lines, others join to form islands covered with mangroves and dry at low water . . . into this river the drains of the most populous part of the town are supposed to empty themselves."[8] Like many colonial settlements of the era, George Town and then Singapore each began its urban life as a defensible fort and harbor, around which grew a commercial area and residences. The urban core was arranged on a grid pattern, with straight roads demarcating plots earmarked for development.[9] Surrounding this were the coastal and riverine kampong (small villages) that predated British administration. For Singapore, in 1822, Stamford Raffles—then lieutenant governor of Bencoolen—had asked a consortium of men comprising Nathaniel Wallich (the notable botanist), Dr. James Lumsdaine (a civil surgeon), and Captain Francis Salmond (Bencoolen's harbor master) to form a special committee, its purpose to investigate expanding the fort and harbor area at the mouth of the Singapore River into a new commercial and business district.[10] The subsequent report, along with input from Raffles, became the basis of a town plan that was drafted by surveyor and engineer Lieutenant Philip Jackson in 1823. The plan required extensive reclamation work on the southern banks of the Singapore River, and the committee noted "there can be no doubt . . . that on raising it and draining it [the south bank] and, by defending it by means of a strong piece of masonry along the river side, not only will those extensive grounds be rendered permanently dry and salubrious but they will become available for building and cultivation."[11] Essentially, this meant trying to prevent the incoming tide and filling in the wetlands with earth. Another of the report's recommendations was that some of the traditional communities living around the river in *roomah rackits* (raft houses) be resettled to make way for merchants, which irrevocably changed a way of life for many by removing settled inhabitants in favor of incoming commercialists.[12] This early reclamation, funded reluctantly by the British government in India, was completed under John Crawfurd after he took up the position as British Resident of Singapore and became the area known today as Boat Quay.[13] This, in combination with the area now known as Raffles Place, quickly became the center of the commercial district, the hub of the Straits Settlements' trading and business interests.[14]

Seawalls and embankments were the key thrust of early coastal flood mitigation strategies but were not without their problems. In 1829, for example, one of the earliest structures at the Singapore River partially collapsed. This was due in part to a combination of shoddy workmanship, nonadherence to the original design specifications, and poor or unsuitable materials. In particular, the mortar used was not water resistant and therefore quickly washed away from the lower sections of the wall. The posts were neither dug deep enough nor placed correctly to afford proper support, and metaphorical corners were cut by using nonstandardized building materials. The committee members appointed to investigate the wall's subsequent partial collapse realized that eight hundred Straits dollars more would need to be spent to rectify the problems.[15] This situation was typical of many such early town planning initiatives.

With the steady transition of sail to steamships from mid-century and an increase in traffic after the Suez Canal opened in 1869, Singapore's New Harbour and Victoria Dock became more attractive than the original port at the mouth of the Singapore River.[16] New Harbour (renamed Keppel Harbour in 1900) was a desirable location for steamships to berth as coaling was difficult on the shallower and more congested waters of the Singapore River, where high tides or bad weather made it near impossible for small lighters to transfer coal to waiting ships, as was the case also on the less sheltered parts of the shore at Teluk Ayer or at Kampong Malacca.[17]

The Singapore River was by no means at the end of its lifespan, however. A new development that reinforced the riverbanks and sea defenses was built between 1859 and 1864 by the man whose name has been immortalized in the construction's name: municipal engineer George Chancellor Collyer of the Madras Engineers. Collyer Quay was built almost entirely by convict laborers and included a seawall that served a dual function as a landing point for the storage of goods.[18] The work was prolonged and arduous, with construction carried out "but one day a fortnight when the tide was at its lowest ebb, and even then there was a foot of water over the land."[19]

The purpose of the scheme, however, was less making the area habitable than making it profitable, and the entire work was funded by local merchants with a vested interest in the quay.[20] This reclamation and seawall were expanded during the last three decades of the nineteenth century in order to build a new seawall across from the commercial center to the (literally named) New Harbour, which was built at Tanjong Pagar across from, and using, the hills of Telok Ayer as landfill material.[21] The scheme also enabled several acres of land to be used for building development and for new roads to link the commercial district with the developing dock at Tanjong Pagar (Victoria Dock, 1868) and New Harbour.[22] These developments changed the face of the seafront across this area, creating a clear boundary between land and water that had not existed previously and forcing traditional settlements further afield. As the dockland area moved and expanded further enhancements to seawalls and new reclamation works became necessary by the end of the nineteenth century. These took place under an ambitious scheme that took around thirty years to complete, designed by the tropical engineering specialists Coode, Son and Matthews, and it linked new sections of the coastline around the Tanjong Pagar and (by the time it was finished) Keppel Harbour area.[23]

In George Town, similar schemes were underway, though some of the earliest amounted to little more than sandbanks, with the exception of the area immediately adjacent to Fort Cornwallis, the original British military base on the eastern shore of that island. The state of affairs was made clear when a new commission was given to Captain Thomas Robertson of the Bengal Engineers in 1806 to report on coastal erosion on the north shore. His 1807 report stated:

> When it is considered that George Town is only protected against this very heavy sea, by a Ridge or Bank of Sand only a few feet above the level of high Water & that this ridge is in some places scarcely more than Ten Yards wide, the possibility of a Breach . . . ought to be guarded against by immediate exertion. The Level of ¾ of George Town being scarcely above the rise of the Ordinary Tide, & some parts lower it is evident that if the Sea were to make a Breach on the Northern Shore it would immediately find a passage across the Peninsula & communicate with the Sea on the S.E. inundating more or less the whole of George Town [which] is for a full Mile, low & Swampy the whole would present a most miserable Scene.[24]

Robertson's solution was to use vast granite blocks to create a sturdy base for a new seawall across the northern seafront. Although this was meant to be a temporary solution, the plan still took time to be initiated as it was held up by a council debate over cost and property rights, the latter concerning coastal landowners who had already erected small preventative measures of their own, at their own expense. The work commenced but slowly and in piecemeal fashion. By 1825, there were still large stretches of the coastline that were unprotected even by this provisional workaround that were liable to erosion as well as flooding, but the worst sections were deemed protected.[25]

Within the towns themselves, urban surface water was mitigated through an extensive system of landfill, canals, culverts, drains, and dry ditches that tended to run beside major roads. This system worked for normal rains but was often found wanting during periods of intense rain, with problems compounded by the types of materials used in these and in adjacent construction projects. In the early days, drains and culverts were mostly made from local materials including brick and wood. However, these did not prove durable in the long-term and they failed under the stresses of the tropical climate. Local brick manufacture could also not keep pace, and the shortfall had to be made up by imports from Hong Kong. Eventually, the situation forced a shift to using iron, and in 1876 the respective municipalities began a long slow process of updating pumps and replacing wooden conduits with iron ones followed by a gradual replacement of culverts with the same material, a great deal of which had to be imported from Britain. Roads were surfaced with laterite, if at all.

Laterite was used widely across French and British tropical empires in the nineteenth century because of its local availability. When a road needed resurfacing, fresh materials could be quarried from the roadside in the rural areas.[26] Cheap and readily available it may have been, but its innate porosity meant it became soft and muddy during monsoon rains. This then exacerbated the impact of extreme weather. Roads frequently became impassable, isolating areas of town during floods and making it much harder to get essential aid to those affected. Charles Buckley—longtime inhabitant, trained lawyer, and author of all things Malayan—recalled how, in early December 1855, for example, "Serangoon district [in Singapore] was a vast lake, and communication had to be made by sampans . . . as the tide continued about high water mark for three consecutive days, the rain remained on the low lands and overflowed the roads, to the depth of two feet in places."[27] The local press and municipal reports corroborate Buckley's account, noting, "As the rain still continues our Scavengers and Road-menders will find employment for at least a month in repairing the roads in and near town, we trust, however, the Municipal Council will avail of any labour offering to effect the repairs as promptly as possible."[28]

The municipal councils were, in the absence of specialized departments at that time, responsible for all issues relating to public works, along with many other aspects of town management. Comprised of notable ratepayers (usually wealthy merchants, colonial officials, and plantation owners) the municipal council (in either George Town or Singapore) was a generalist body of members who oversaw the police and (often ad hoc) various committees for lighting, building, and market regulation.[29] They also had oversight of flood alleviation by means of public works, which included road, bridge, tidal gate, and canal building and maintenance.[30]

These municipal governments continued to be rather extemporaneous bodies until 1856 when their powers were formalized and strengthened by the Act of Municipality and their direct line of governance was transferred from the EEIC to the home government. This remained the situation until 1887 when the first municipal ordinance by the Straits Settlements government was passed, and the councils thereby gained more power and autonomy. By the end of the century, local government had begun to develop into more professional and specialized departments such as for public health or sanitation.[31] The 1887 municipal ordinance also created municipal medical departments set up in Singapore, Penang, and Malacca. Although the Medical Department had traditionally worked closely with civil engineers, the new local branches meant that public health really became far more integrated within urban planning. This was a boon as flooding was also an issue of public health as well as of meteorology, creating the circumstances for vector-borne diseases to spread and water supplies to become polluted from overwhelmed drainage systems.[32]

Engineered solutions to prevent or mitigate flooding had been closely connected to ideas about public health, as much as commercial development, from an early stage, far more so than meteorology. Indeed, one of the reasons given for the necessity of the early Boat Quay reclamation was "the health of the Settlement."[33] Nevertheless, there was some difference of opinion on this topic. The report of Wallich, Lumsdaine, and Salmond noted, for example, the "healthy appearance of the Chinese settlers [on the south bank] under various circumstances of exposure and the total absence among them of diseases peculiar to low and marshy situations." This was evidence, they argued, that, despite the common assumption of marshland creating the opportunity for disease to thrive, the area was as "perfectly salubrious" as any in the colony.[34] Their language echoed that of Crawfurd, who likewise noted how "fevers and dysenteries were of very rare occurrence . . . [which] may appear at first view the more remarkable, since a considerable portion of the site of the town . . . is a low and noisome marsh."[35] His contemporary John Caswall, a medical officer of the Madras Medical Service employed in Singapore in the 1820s, described a situation of near "complete saturation" caused by perpetually "moist" airs, and unsanitary swamp "exhalations," yet he was astonished that the colony was in fact remarkably healthy and largely free from "malignant intermittent fevers" and malaria.[36]

Others were less enthusiastic. Drs. Robert Little and T. M. Ward analyzed the incidence of disease and disease types in conjunction with dry, wet, or hot periods, warning of the perilous effects of humidity and excessive rainfall on human health.[37] Little, for example, argued that excess water combined with decomposing animal or vegetable matter was the cause of toxic miasmas, a particular problem in swampy areas where large-scale forest clearances were taking place, as in Singapore and Penang.[38] The onset of the annual northeast monsoon was particularly to be feared, as popular belief held it to herald the "commencement of the sickly season from its excessive dampness."[39]

At that time, the cause of malaria was poorly understood. Instead, marshes were assumed to give off noxious fumes or miasmas that created the malarial fevers which were so dangerous to the early settlers. Swamp drainage was therefore a priority, and urban drainage systems and sewers absolutely had to be fit for coping with heavy, as well as normal levels of rainfall, so that they would not become overwhelmed and contribute to the spread of disease or create pools of standing water. Important here were the rainfall records made by the Medical Department, which had become more readily available from the 1880s, especially for the urban centers of Singapore and George Town. These records were used by engineers to predict worst case scenarios and their frequency. In George Town for instance, the main channels for diverting floodwaters were "constructed across the general line of drainage . . . arranged as to intercept the flood water flowing from the suburbs towards the town and lead it by shorter routes to the sea" using a combination of earth, concrete, and brick. Moreover, main drains were built to have capacity to discharge one inch of rainfall per hour, a figure found to be "about sufficient for heavy rainstorms."[40]

As this above account, based on discussions with Municipal Engineer of Penang L. M. Bell hints, the urban core was prioritized over the suburban areas. In the outer districts, it was not considered necessary to incur the expense of constructing channels that could entirely prevent flooding simply to be sufficient to carry off water within twenty-four hours.[41] Later, when the vector-borne theory of disease became common currency in the early twentieth century, sanitary engineers also focused on joined-up solutions, to kill mosquito larvae as well as to improve drainage works so as to prevent standing water. However, given that there was no "self-cleansing system" in place at that time and such drains relied on finding "low-caste Indian coolies" to undertake the work of sweeping and flushing out the main drains and channels, one wonders to what extent the workers were protected from mosquito-borne disease.[42] When, in 1910, a new branch of the Medical Department was created specifically to oversee Straits Settlements public health, leading to the creation of an "Anti-Malaria Committee" in 1911 under its auspices, public works to prevent malaria began to rely more on chemical solutions such as limewashing, the physical destruction of breeding grounds, and the draining of swamplands.[43] This permanent committee's membership included the municipal engineer whose role overlapped with those of the new health officers who worked together for malaria prevention, fresh water, and sanitation.[44]

INTENTION VERSUS REALITY

For much of the nineteenth century, flood control tended to be ad hoc and reactive despite the advent of many large improvement schemes, a situation not dissimilar to the rest of the British Empire. In the words of Colonel George Chesney on the Indian situation, the "Court of Directors . . . until almost the termination of their existence, did not recognize the prosecution of public works as a necessary part of their policy."[45] His words were broadly applicable to the Straits Settlements though they did not take into account the local municipality's concern to improve the situation. What they do reveal, however, is one of the reasons that the municipality could not do more, despite their best intentions.

During a flooding episode in George Town in 1847, there was "havoc . . . on the roads," as the *Singapore Free Press and Mercantile Adviser* noted: The damage left for "public works is . . . considerable. . . . Almost every road is severely cut up. The surface, at different parts—for many yards in length . . . is almost completely washed away, leaving the ragged, stony foundation exposed. In the neighbourhood of town, the Pinang and Macallister roads—the widest and best kept in the island—have suffered especially." The costs, estimated at thousands of Straits dollars, were considered overwhelming, and the municipality had no access to the sums of money required.[46] One of the major issues was the way in which revenue was raised. The EEIC and British government assumed that locally raised taxes and revenues would fund all but the most major engineering projects or that the colony could rely on the generosity of wealthy local philanthropists. After all, people were getting rich from cheap land, plantations, and trade.[47]

Proceedings at both Singapore's and Penang's municipal councils also shed some light on how funding worked and the extent of the problems. Everyday municipal works were funded by fines levied by the Magistrate's Court for urban offenses or from rates chargeable on property, but in the early years there were few rateable properties (criteria included whether the property was made from brick) and there was much resistance to paying.[48]

To make things cheaper, the municipalities utilized the extensive supply of free convict labor for construction works. It is worth explaining here that around this time there were just under two and a half thousand prisoners incarcerated in Singapore, "the majority of whom came from Bengal and Madras. These are employed in making roads, assembly bridges and digging canals; and undoubtedly without them the town . . . would have been now but a sorry residence."[49] Were it not for the ability to traction on low-cost manpower, many municipal schemes, improvements, or repairs would have been in danger, including the provision of fresh water infrastructures.[50] In Penang after the 1856 floods, for instance, it was convicts who maintained "all the roads and bridges . . . after the damage effected by heavy rains."[51] However, although cheap labor was readily available, materials were not. "The principal drawback to maintaining in good order the numerous and extensive highways," it was noted in 1856, "is occasioned by the extreme difficulty in obtaining suitable materials, except from a considerable distance, and also the fact that some of the chief roads in the country having been made through swampy ground and filled in with fascines. . . . it will be requisite to incur a large outlay in reconstructing some of the roads in question."[52] Labor was cheap, but the bugbear was, and would continue to be, the availability of money and resources.

Writers of open letters to the local press acknowledged the lack of funding and resources, but they also pointed to other problems. Long lists of municipal improvements can be seen documented at every meeting of the legislative council. Items included increasing the number of culverts, widening, deepening, and improving canals, remetaling roads, and rebuilding wooden bridges with brick and iron.[53] Nevertheless, the mounting number of incidences of flooding recounted in the local press reveal how these improvements were not resolving the difficulties. During severe weather in 1891 and 1892, for example, residents complained that canals were not fit for purpose, culverts were poorly maintained and frequently blocked, and floodgates were left closed when they should have been opened.[54] The Singapore River, too, provided a continued source of grievance with seemingly no resolution. It was noted in 1884 that the practice "of refuse and filth being thrown into the river from the numerous boats and from the quays" had led to shoaling within the river basin and hence contributed to the scale and frequency of river floods, leading the municipality to commission the eminent civil engineer John Coode to design a river improvement scheme. Forty years later, however, the scheme was still being debated by the municipal council.[55]

Much of this was undoubtedly due to lack of resources, but the municipality's ability to manage floods, at least until the 1920s, was further disadvantaged by ambiguity over who did what. Canal administration is an excellent example of this. The Bukit Timah Canal, Rochor Canal, and the

Singapore Canal were Singapore's main urban canals. Their purpose was to provide the means for excess floodwaters to flow out of the town center. As sections of these waterways were initially constructed within the harbor limits for the purpose of facilitating the passage of small vessels, they fell under the government's jurisdiction, not that of the municipal council.[56] Unfortunately, although they were clearly valuable to maritime trade and industry, the government had little interest in their day-to-day maintenance. Thus the free flow of water was frequently hindered by the buildup of rubbish and silt; even floating logs generated by adjacent sawmills contributed to flooding during heavy rains.[57] Although some of these problems were alleviated by a further transfer of control from government to municipality in 1919, there was a similar snagging point at the boundary of town and rural jurisdictions.[58] The municipality only had authority over the urban core, a division that had been established in the nineteenth century when the town was small and relatively free from sprawl. As the population increased by around one hundred thousand inhabitants every decade from the 1880s onward, the town developed rapidly, much outside of the original demarcation.[59] For example, in 1918, the drainage channel that ran along the municipal boundary from Geylang Road to the sea, acting as the main drain for what was then an agricultural district called Payar Lebar, became blocked but, with neither the rural board nor the municipality claiming responsibility, the neglected canal caused serious flooding during the southwest monsoon season in the coastal region.[60]

Such incidents had important implications for urban flood control because of their cascading impacts, even though their location was outside the immediate town limits. Flood alleviation works outside the town limits were minimal. For example, the municipality only dredged the section of the Rochor Canal from Kandang Kerbau to Victoria Bridge. The other section of the canal, leading from Victoria Bridge to the sea, fell outside municipal boundaries and was therefore not managed. As it gradually silted up, the reduced water force reduced the canal's capacity to carry away floodwaters from the town center. By the 1920s, this and other similar situations had led to an urgent call to improve cooperation between the municipality and the rural board to improve river and drain management, especially through increasing dredging, by straightening water courses, and by raising bridges.[61]

Instead of improving, though, things became worse from the 1910s and into the 1930s, despite more investment in improvement schemes. During the 1904 discussion of George Town's municipal budget, for example, the municipal government had put forward a fourteen-thousand-dollar budget "to avoid the flooding" by building a new bridge and a wall round the Pig Market in order to improve current obstruction to the flow of the drainage waterways.[62] The municipality also undertook measures to improve the

Sungai Penang river—"the principal flood outlet of the Town"—by purchasing a new dredger in 1908.[63] Nevertheless, these measures seemed to provide temporary stopgaps rather than real solutions, with the municipality officials unwilling, or unable, to make a sustained investment in flood alleviation works, especially for the Penang River, which was still silting up and causing floods for decades to come.[64]

Indeed, across the peninsula during the interwar era, there was a series of severe floods—beyond the monsoonal norm—that entailed serious outcomes on the urban environment and the people of the settlements. During Christmas 1926, for example, unusually heavy rains resulted in record-breaking floodwaters across Malaya.[65] In Kuala Lumpur, the Rev. D. Fergus Ferguson of St. Andrews Presbyterian Church described how there was a "metre of water in the centre of K. L. Office and shop workers paddled to work in sampans and, when the waters receded, the Chartered Bank spread millions of soggy bank notes to dry on the Selangor Club's *padang* [field]."[66] A day later, an observer told how water had "begun to overflow the streets. The *padang*, which was clear this morning is once again under water and on all sides the water is rapidly rising." In some places, water was six feet deep, making communications "between one portion of the town and the other . . . totally impossible." Another five-foot-deep "lake" had formed, which cut off the Empire Hotel from the Klang River, and at "Old Market Square and the business offices, the water was over four feet deep."[67]

In 1933, too, the wealthy Tanglin residential area of Singapore was once again cut off from the main town by waist-deep water in the northern part of Orchard Road as heavy rains combined with tidal flooding, and then, a month later, exactly the same thing happened again.[68] For the most part, the government was blamed for poor management that had contributed to making these events worse. "Contrary to all precedent . . . five severe floods have occurred through the overflowing of the Bukit Timah Canal," wrote a disgruntled member of the public about floods that had taken place in 1928. "For a long time past, the canal has been silting up . . . the water from the canal is obstructed at the sea end, and cannot get out . . . it is quite clear that a rainfall of only 2½ inches is quite sufficient to flood the whole neighbourhood."[69] Then, in 1931, it was noted how "despite many improvements, flood prevention work in Singapore is by no means completed, and the flooding which occurred in the Cuppage Road and in Balmoral Road district following the heavy rainfall on Tuesday afternoon, shows that the position in regard to these districts at all events . . . remains unsatisfactory."[70] One year on, in an open letter to the *Straits Times*, "A Ratepayer" noted how the Bukit Timah Road was flooded again, damaging many properties: "I have lived in this locality for the last seven years . . . and as far as I know no effort has been made by the municipal engineers to improve matters, but instead floods have been more frequent."[71] As another writer also angrily noted after yet another incident in January 1934, "it has to be admitted that the rainfall during the weekend was abnormal; but abnormal rainfall is a periodical occurrence in Singapore."[72] The list of complaints was far longer than this short list of examples shows, but these serve to illustrate something of the scale of the problem that reached across urban, peri-urban, and rural areas, creating an expensive headache for the British administration across the Straits Settlements and the FMS.

The worsening issues across Malaya led government officials to convene a special investigative committee in 1927. Presenting their first report one year later, committee members observed that environmental degradation was the chief cause of the worsening problem, in large part because of excessive siltation from soil wash and erosion caused by mining, agricultural activities, and deforestation.[73] The scale of the problem was such that even the American consul to Malaya penned a special report on the subject to the US secretary of state, no doubt because of American dependence on the local rubber industry (rubber plantations were frequently affected by flooding). "In reference . . . to the serious matter of flood control in Malaya," the consul wrote, "its seriousness is increasing with industrial development. . . . Yearly, the many streams in Malaya have, during the so-called winter months, overflowed their banks causing serious damage to property and life. . . . At times entire villages have been completely wiped out, rubber estates have been destroyed and other damages following flood seasons have been apparent. This situation which has been a perennial subject of discussion, eventually reached a point which called for some definite action on the part of the Government."[74]

It was also quite apparent to district officers working in parts of the FMS that flooding ought not to be viewed in a silo and that preventative solutions had to consider industrialization, infrastructure, soil erosion, deforestation, and land-use change in combination with mitigative engineering works.[75] This had been the conclusion of C. E. Dupuis, a leading British engineer appointed as the colony's advisor on irrigation and flood prevention, a new post, during the same decade. Dupuis warned of how "the recent and rapid development of the country" was exacerbating degenerative cycles, whereby roads and towns had been constructed in

> unsuitable positions without adequate provision for floods . . . the natural conditions have been greatly changed and the flow of the rivers gravely affected. This has been caused mainly by tin mining and by the clearance of natural forests over large areas and their replacement by rubber or other plantations . . . the conversion of dense tangled vegetation into orderly plantations with clean weeded ground surfaces has enormously increased the

> rapidity with which the water is discharged into adjacent rivers . . . and the escape of large quantities of course sand from the mine working into rivers has choked and destroyed their channel over considerable lengths.[76]

In 1934, government adopted a five-year plan to address flooding across the whole peninsula, with targeted measures for trouble spots where flooding repeatedly occurred. In Singapore, the plan included widening and deepening the Bukit Timah and Orchard Road Canals to allow for better channeling of water away from the perennially flood-prone Bukit Timah, Orchard Road, and Whampoa areas.[77] Other work was carried out in conjunction with the new Drainage and Irrigation Department, which served the whole colony and was intended to be a more joined up solution to combat the related issues of soil erosion and riverine siltation, among other things. At George Town in 1932, for instance, heavy rains overwhelmed the banks of the Sungai Penang, which had silted up almost entirely, flooding adjacent land. The five-year plan was designed to tackle this in conjunction with the new department, which employed eighty-three manual laborers ("coolies") to remove fifty-three thousand cubic yards of silt from seven miles of river by 1938.[78] This was only the first stage in alleviating the problem as the Sungai Penang was still flooding annually into 1939, but it marked a significant improvement from what had gone before.[79]

EXPLAINING THE INTENSIFICATION OF STRATEGY

One may wonder why it took until the 1930s before government officials embarked on this series of coordinated flood responses. This was a decade of acute economic stress and monies for works would have been tight. There may have been two possible reasons or strategies at play to explain this. First, the British government—whether in London or in the Straits Settlements—rarely moved quickly on large infrastructural schemes. Early nineteenth-century coastal defenses in Penang had been delayed because of debates over ownership and funding and the Singapore River improvement scheme took more than forty years from the 1880s into the twentieth century to reach its apex.[80] Another scheme, intended to introduce a much-needed waterworks reservoir (now MacRitchie Reservoir) took over twenty years from the initial philanthropic donation to its completion.[81] Given the elephantine pace at which the government moved on such things, it is perhaps no surprise to realize that the schemes that came to fruition in the 1930s had, in many cases, been vaunted and agreed upon in the 1920s or even earlier, when money was less of a concern.

The early 1920s saw a relatively strong economy—with canned pineapple, tin, and rubber all seeing profitable margins, alongside a new commercial tea industry breaking into local markets from 1927.[82] Rubber was sovereign of all the produce, however. Even in 1929, just before the Great Depression, Malayan GDP was higher per head than in any other Asian country, the direct result of its being the world's largest supplier of rubber.[83] It comes as no surprise that the local planting community had a strong voice in government and policy. Many plantation companies also appointed agency houses to manage the estates on their behalf. These agencies could be very powerful, often controlling multiple estates and undertaking large investments on the London stock market. Some of the largest such multinationals—Sime Darby, Guthrie, Bousted, Harrisons, and Cumberbatch for example—held substantial power and influence in the City of London.[84] These agencies and the planting associations had the ear of officials in both local and home governments managing the instabilities of the market during the 1920s and 1930s—providing input into top-down schemes to raise prices by limiting production such as the short-lived Stevenson Rubber Restriction Scheme of 1922 and the International Rubber Restriction Scheme that was adopted from 1934 until 1942.[85] The United Planting Association of Malaya was deeply invested in pushing for joined-up solutions to improve productivity. There was, for example, strong support among the planters for the establishment of a Rubber Research Institute (RRI) to further exploratory research in the field, including soil research.[86] Although it took a few years to persuade government officials of the institutes' value, permission for the facility was eventually granted in 1925 on the basis that the investment would facilitate returns.[87] The United Planting Association was involved in other initiatives too, including membership on a cross-organizational standing central committee to investigate soil erosion, which was established in 1939.[88] All these organizations had a vested interest in combating flooding as a preventative against further soil erosion and loss of crops and soil fertility.

Investment in flood mitigation, then, was a way to aid productivity as much as to improve public health. This linked well with contemporary government strategy to implement and promote higher levels of internal self-sufficiency and to rely less on global markets, especially after the Malayan economy worsened in 1931. In response to trade deficits between 1930 and 1932, local government officials looked internally to use local resources including food and labor more effectively, rather than relying on traditional imperial connections—a strategy that, Paul Kratoska has argued, enabled Malaya to weather the worst of the financial storm and emerge more self-supporting and less dependent on imperial networks at the end of the 1930s.[89]

In line with this shift in strategy (but also a product of the times) was the diversification and specialization of administrative departments. In 1932, for example, a new, specialized Drainage and Irrigation Department was established. Originally the Hydraulic Branch of the Public Works Department, this new administrative iteration was formed on the recommendation of the Rice Cultivation Committee as a means of encouraging greater rice security within Malaya. It never really lived up to this purpose as, although rice production did increase for a few years, a small recovery in rubber prices during the mid-1930s led to farmers' continuing to prioritize rubber over rice.[90] Instead, the committee played a critical role in advising on soil erosion and flood prevention, working with other concerned departments including Lands, Mines, and the Agricultural Department. Thus, improving the infrastructures for flood prevention worked hand in hand with other areas deemed essential to colonial interests.

From the 1920s too, public health and housing reforms took on an added intensity. Although these were not directly aimed at flood mitigation, reforms impacted on living conditions, overcrowding, and urban health. A new public health branch of the Medical Department was created in 1910. As a result, each settlement gained a municipal health officer whose remit included malaria prevention, fresh water, and sanitation, all of which entailed working with engineers on draining and filling in the swampy land that provided breeding grounds for mosquitos across urban and rural areas (especially at plantations) and improving urban drains to prevent houses from becoming damp and unhealthy.[91] A report created in 1906 on the sanitary condition of Singapore, for example, noted how shophouses—a terrace-style working and living space—had inadequate drainage systems and oftentimes extra rooms created beneath the original ground floor, leading to permanently damp, easy-to-flood, unhealthy living spaces with high levels of disease.[92] This was one of the many reasons for the establishment of the Singapore Improvement Trust in 1927 under the Singapore Improvement Ordinance, with an initial agenda to improve drainage and to earmark unsanitary buildings for demolition.[93]

CHANGING LANDSCAPES AND SOCIAL VULNERABILITY

This points to an important sociocultural implication as to who was most affected by the floods and how government responded. To explain this more fully, the narrative will take a brief diversion. Across the 1930s, the acceleration of motorcar ownership demanded better roads. Surfacing materials transitioned from laterite to the more expensive but more durable "metaled" surface, comprising granite, limestone, and bitumen. While roads became more flood proof, cars did not. Heavy rains and flash floods caused cars to stall; they flooded engines, and somewhat ironically, people trying to get to work on rainy days often found themselves abandoning their cars and resorting to the more water-resistant rickshaw puller instead to get there.[94] The car users were colonial officials using office cars, or wealthy inhabitants

FIG. 2. "Plan of Singapore Town Showing Topographical Detail and Municipal Numbers," 1893. Sheet 3, 20140000077-0196_SP002988, NAS.

who owned property outside of the immediate town center. The suburban bungalow areas of Tanglin and River Valley that developed in the later nineteenth century are good examples, shown in detail on a town plan of 1893, marked with gardens and private driveways.

Averaging around twenty-five or thirty meters above sea level, this area may have been a modest elevation by most standards, but it was nonetheless an improvement of at least fifteen meters on the districts abutting the river, the sea, and the swamps, which were (unsurprisingly) prone to flooding.[95] These bungalow residents complained frequently of being cut off from the town during floods but their areas were rarely flooded themselves. That this differentiation was obvious to contemporaries can be seen in popular culture and jokes of the time. In 1933, for instance, the *Straits Times* printed a dialogue that poked fun at the towkays (usually wealthy businessmen of Chinese origin): "Two towkays met in Raffles Place and one told the other that he had insured his new house for $50,000 against fire or burglary. 'But haven't you insured against a flood?' asked his friend. The owner paused for a while, and then said, 'How do you make a flood?'"[96]

Straits Settlements society had always been characterized by a wealthy minority and a poor majority with some of the steepest population disparities in the early nineteenth century.[97] During the interwar years, however, despite an exodus of plantation workers overseas as opportunities dried up, the population continued to rise, especially in the towns. In Singapore, this meant roughly doubling the population between 1901 and 1931.[98] It was immigration, not the birth rate that was driving the population up, and the new settlers had no choice but to live in the cheaper areas. As was, and is, the case in many towns and cities, this meant the lower-lying, flood-prone riverine areas and swamplands. The houses in which these migrants found themselves living also tended to be overcrowded and, unlike some of the native villages that were built to withstand the ebb and flow of water, were poorly constructed from cheap materials with little consideration for the environment.[99] In George Town, there is ample evidence of teeming squatter settlements in and around the low-lying flood-prone banks of the Penang River where the very poorest lived on the ground floor and were vulnerable to every flooding event and the associated hazards of ill-health and disease.[100]

For Singapore, John Turnbull Thomson's detailed 1846 survey reveals the extent of the marshland across which large portions of Singapore were later developed. The south bank of the Singapore River in the 1840s, for example, "was some nine feet lower than the opposite creating, at high tide, a vast inland lake up to 12 feet deep that engulfed the entire Chinatown area to Pearl's Hill, and at low tide a marshy bog."[101] On the map, Pearl's Hill is specifically marked. Chinatown is not marked per se but it is bordered by three "Chinese Burying grounds," to the west of the town core (see figure 3).

Although engineering had eased some of the environmental triggers by the 1930s, social disparities and a lack of resilience to flooding characterized the life experience for many of these migrant rich spaces and the new brick, stone, and wood towns were entirely reliant on effective environmental controls to mitigate the effects of excess water.

For the traditional communities of people who had inhabited the island prior to the British arrival and for the many small rural kampong that grew up following Chinese immigration, the situation was a little different. For the indigenous communities of the rivers, marshes, and peripheral agricultural regions, minor monsoonal or tidal flooding was a normal part of life. The traditional kampong of the southern coastline were predominantly comprised of Malay, Bugis, or Orang Laut peoples whose livelihoods depended on the water. In a descriptive account of the Orang Laut, for example, made by John Crawfurd, a physician who later became British Resident at Singapore, these fishing communities used the incoming flood tide as a means of easily moving boats from the shore to the middle of the river channel in tune with the natural movement of the water. Here, they could spear or net the

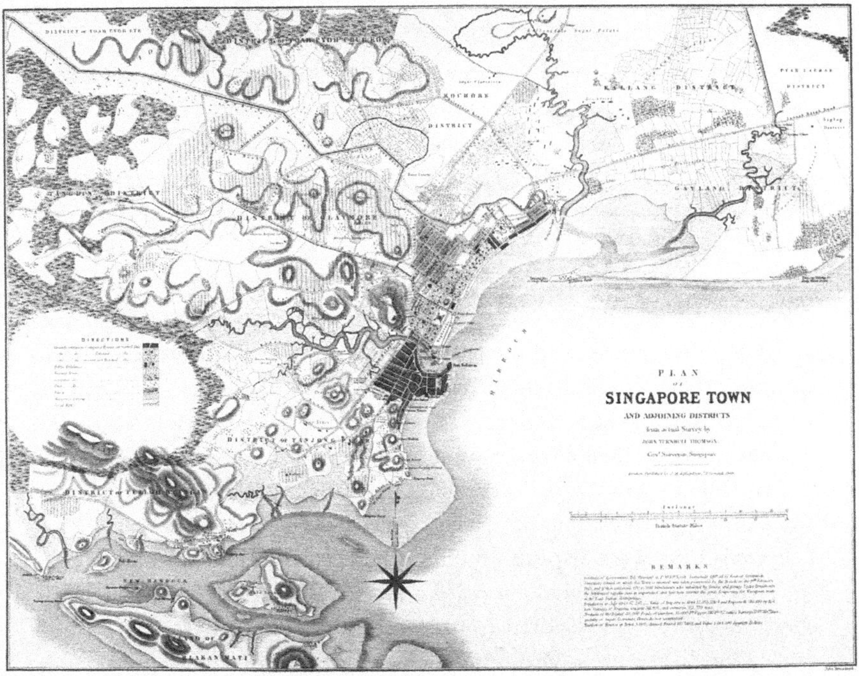

FIG. 3. "Plan of Singapore Town and Adjoining Districts from Actual Survey by John Turnbull Thomson, Government Surveyor, Singapore," 1846. Media image no. 20080000141, accession no. 3809a, NAS.

abundant deeper water fish.[102] In 1851 Robert Little wrote that "hundreds of houses" inhabited by "both poor and comparatively wealthy" Malay families also dotted the banks of the river, made of "wood, or *attaps*, [and] erected on posts." Although the swamp "is only one open flooring of *Nibong* [palm] away, the elevation of these houses is just sufficient to place them above the high water mark."[103] These dwellings were well adapted to the constant flows of the tide, monsoon, and flood.[104] As new migrants arrived, however, these communities were forced to the margins, and their ways of living and managing the water were overwhelmed by Europeans' supposed technological superiority and control of trade in the region, with the Orang Laut, in particular, becoming, as Timothy P. Barnard notes, the object of "government scorn and anthropological enquiry."[105]

Many of the Chinese migrants who arrived in Singapore following the advent of British rule, ran small-scale farms that developed in the eastern part of Singapore, especially at Lavender, Bugis, Kallang, and later, further east at Geylang and Paya Lebar. These were more in tune with their environment, working with what was there to produce vegetables that thrived in the conditions, such as sweet potato, tapioca, lettuce, onions, celery, cucumber, radish, brinjal, bitter gourd, and soya beans for local markets.[106] Many of these villages were largely self-governing, operating within their own hierarchical systems of community elders or clan associations. With the spread of urbanization, many of these former communities broke down as the government forcibly resettled them to make way for land development and reclamation projects in the twentieth century, impacting on the uprooted communities' resilience to flood events.

In the early 1950s, for example, the districts of Paya Lebar and Bedok to the east of Singapore town were a largely rural area with two small coastal fishing communities and some small villages mainly comprised of ethnic Chinese. Their second- or third-generation farms provided fresh vegetables, pork, chicken, and eggs to Singapore's town. The low-lying land was an advantage under normal circumstances as the wet ground was well irrigated and suitable for growing certain crops, including types of Chinese cabbage and *kang kong* (water spinach). During 1953 and 1954, however, many Paya Lebar residents were displaced by the British government to a new site on the coast at Bedok, to make way for a new airfield development. Many of these residents were removed from their traditional support networks of community and clan associations and were in the process of starting afresh when a series of devastating floods occurred during the last weeks of 1954.[107] The Bedok Flood Committee—a new organization established in response to the events—raised how poor living standards, the unsuitability of the new site, and the rising cost of living all contributed to residents' vulnerability to the disaster. "Many of us are now in debt," their official letter read to the government read, and "we cannot hope to get back, at any time in the foreseeable future, to the financial position we were in at Paya Lebar."[108]

The case is a good example of how public expectations of government had changed by the twentieth century, especially following the Second World War. While a large proportion of complaints to the papers were penned by middle-class ratepayers (often European or Eurasian), shifting political ideas on social reform and welfare across the spectrum of liberal, conservative, and socialist thought in the early twentieth century, especially during the interwar years, were also changing people's expectations of government at the same time as increasing state intervention in the community. In the British context, Gregory Claeys has argued how the move "from a suspicion of the state as essentially the agency of corrupt aristocratic government to a concept of it acting as a moral and ethical arbiter among, and promoter of the welfare of, all classes" remains the most significant conceptual change of the era.[109] As the nineteenth-century thinker Thomas Carlyle noted, this shift resulted in increased government regulation as a form of social duty by the landed (political) class in the public sphere, especially in the shape of civic legislation.

By the early twentieth century, political thought had changed significantly with the rise of socialism and the extension of the franchise in Britain, Europe, and the United States. In the Straits Settlements, calls for wider social representation in politics and self-government were apparent during the interwar period, many of which were led by the local elites who had been schooled in Britain and were directly exposed to this new, radical landscape. Immediately following the war, calls for independence amplified, and tensions erupted in the form of a supposed communist uprising during the late 1940s and 1950s, which the British government supressed with force, known as the Malayan Emergency. Events such as floods, which exposed government failings, directly impinged on the well-being of local communities and were easily exploited in critiquing the government and with good reason. The Bedok floods, for example, were appropriated by supporters of the new Labour-Front government that came to power in 1955, as a critique of the slapdash handling of flood management and blinkered treatment of more vulnerable social groups by the colonial government.[110] Unlike the nineteenth century where the British ports of Malaya were little better than frontier towns, by the twentieth century they were mature societies. Expectations were correspondingly higher for what government could, or should, provide.

A CHANGED LANDSCAPE

Looking back with sentimental hindsight over the developments of the early twentieth century, Singapore inhabitant Roland Braddell eulogized the mastery of humans over nature: "I watched Mount Erskine and Mount Palmer and many other small hills being hacked and hewed and spewed into the swamps, great areas of reclamation created, the tide beaten back foot by foot to its present race, great walls of granite rising forty feet from the concreted bed of dried-out swamp, the sea brooked and curbed and shut out by more granite walls, until to-day all trace of the great swamp has gone. . . . I rub my eyes with astonishment at what I saw done in those six years between 1908 and 1914."[111] Braddell's statement may have exaggerated the actual situation, but it is indicative of the many changes to Singapore's natural landscape and infrastructure over the early twentieth century. The five-year plan designed to combat the challenge of repeated inundations, developments across all areas and more joined-up approaches to tackle flooding and drainage, erosion, and soil productivity, as well as a big push to improve public health in urban areas through swamp drainage (inspired by changing knowledge of vector-borne diseases such as malaria), all combined to bring about an improved situation by the eve of the Second World War.[112] Many of these improvements combined multiple fields of science in addition to engineering, and some of these sciences were comparatively new to the twentieth century. Much of the work was undertaken in Malaya in the proliferation of new specialist departments and experimental stations that had grown up since the 1920s, including the Department of Agriculture (1903), especially its soils division, and the Drainage and Irrigation Department (1932). Many of these department officials undertook their own work in meteorological observation to assess scale and need. However, the authorities saw floods as part of a needs-based agenda for engineered mitigation rather than as a meteorological issue per se, unlike the more closely investigated droughts of the following section. Engineered solutions had to take into account predicated rainfall and potential flood heights, but the reasons for and the predictability of flooding were more of an engineer's concern than a meteorologist's.

These changes were by no means a panacea (extreme events such as that of 1954–1955 are a case in point), but government officials were taking steps to better manage the problem. To see the colonial government as being in control of the situation, however, would be a step too far. A sense of continually "catching up" is more evident than a sense of control in the many schemes that were implemented, and this is not necessarily a rebuke. As many scholars of disaster argue, control in the face of extremes was, and is, not fully possible even with our modern technologies, only risk reduction is.[113] The question, then, must be one of assessing the colonial government's success in mitigating, rather than preventing, disaster.

CHAPTER 6

DEALING WITH DROUGHT

AN ARTIFICIAL RELATIONSHIP WITH WATER

By the early 1900s Malayan government officials had at their disposal rainfall records that stretched across several decades, courtesy of the Medical Department, and there was now strong evidence that droughts had a periodic frequency, but the weather science and the engineered solutions to mitigate such events were not aligned. Droughts constituted a serious problem for the tropical colony. They threatened agricultural productivity, endangered trade, and imperiled the health of urban inhabitants. With some irony, the more economically successful Malaya became, the more stress was placed on extant water resources, as town populations rose and commercial plantations, emerging industries, and growing port trade demanded ever more water. Engineered solutions to provide a constant water supply—whether for towns and commercial shipping or for agriculture and plantations—were available, so the problem of water supply was at its heart both political and economic.

Britain's experience in India witnessed the birth of modern Western-style hydraulic engineering on a grand scale; from adapting indigenous watercourses and dams or by imposing new forms of irrigation systems, canals, dams, and reservoirs as a means of mastering nature.[1] Yet, despite this readily accessible knowledge and availability of a new class of technocrats and skilled experts in engineering and engineered solutions, officials' ability to manage water in Malaya was limited by political will, and thus the funding and resources (including skilled manpower), needed to make the colony water resilient. Deep social inequalities also meant that access was uneven to whatever water supply systems were available. Indeed, the types of land transformation that colonialism engendered (and especially urbanization) had the consequence of changing the relationship with water for some social groups. This phenomenon has been studied extensively in an Indian context, where colonial hydrological engineering works ushered in an era of technological lock-in, and the views and solutions of the colonial technocrat were not always suited to local environmental and climatic conditions.[2] Exploring the progression of, and approaches to, hydraulic management and thinking about water security in Malaya, this chapter claims that the experience of managing drought and ensuring water security can be viewed as an expression of governmental priorities, and the extent of their control over the local water environment can be seen as a metaphor for the complexities of colonial bureaucracy and of engrained attitudes toward the people of the tropical region. Colonial officials' lack of reactivity to the climate, therefore, and despite new knowledge on rainfall challenged the day-to-day experience of weather under colonial rule.

WATER SECURITY IN EARLY NINETEENTH-CENTURY SINGAPORE AND GEORGE TOWN

One of the earliest water reserves in Singapore—a tank to supply fresh water to ships docked at port—was installed at the foot of Fort Canning during the 1820s.[3] Already by the 1830s, it was obvious that this small works was not sufficient for the port, let alone the town, not simply because of everyday needs (though this was a problem) but because the tank was not replenished during periodic dry spells. The municipal council made monies available to provide for a backup water tank to augment the Fort Canning facility, but it was not the government but instead the generosity of inhabitants that made the first real inroad into creating a reliable, portable water supply for the town. A significant sum of money was donated in the 1850s by municipal council member and philanthropist Tan Kim Seng to build a large reservoir and waterworks, which the town urgently needed. Nevertheless, political considerations, construction issues, and additional financial costs meant that the project took just over twenty years to complete.[4] It was meant to open in 1867, but a major error in calculation resulted in the works not being built at a sufficient height to provide the necessary pressure to drive water to the town.[5] Thus more monies and another decade were added to its making. The facility—originally known as the Impounding Reservoir and still in operation today as MacRitchie Reservoir—was constructed on the gravitation principle, a British supply system commonly adopted in Britain's Asian colonies.[6]

In George Town, too, the situation was much the same. One of the earliest works was a brick aqueduct designed to channel water from a waterfall at Mount Olivia to the town, a distance of some six miles, commissioned by then governor Robert Townsend Farquhar in 1804. Like the early storage tank at Fort Canning in Singapore, the aqueduct supply in George Town was designed to facilitate water for trading vessels ships docked in the harbor, though it also connected some domestic and civic buildings through a system of earthenware pipes.[7] As the population of George Town more than doubled in the subsequent two decades, the aqueduct had to be rebuilt in the 1830s and was later extended east into the coastal area of Pulau Tikus (incorporating the site of the Penang Botanic Gardens). This made some inroads into providing water for the settlement until a reservoir and waterworks were constructed at the westerly hills at Ayer Hitam, which were later overhauled in the 1890s.[8]

Singapore's long-awaited reservoir eventually become operational at the end of 1877. Ironically, this was the year of one of the worst droughts then known on the peninsula, and as might be imagined, during the height of the drought the torpid reservoir became a symbol of governmental incompetence and malaise. This terrible drought resulted in severe restrictions on drinking water, including the length of time that standpipes could operate and the closure of many wells because of contamination. The standpipes and wells were normally fed by underground tanks and small reservoirs—supplemented in some cases by river water or by natural springs located on the town's outskirts or in adjacent villages—and were for many urban inhabitants the only source of water. It was not long before cholera was reported as result of drinking contaminated water in the poorer suburbs of the Straits Settlements' towns and their adjacent kampong.[9]

This event, just like other severe events of the nineteenth century that took place in 1864, 1885, and 1895, had a major impact on food security.[10] Malaya produced many foodstuffs commercially, but few of them found a home in local markets, with the exception of some fruits and vegetables grown around or near settlements and industrial areas to service immediate needs. The staple diet for most people was rice, and although several Malayan districts were rice-producing, not enough was grown to supply the domestic market.[11] For much of its rice needs, therefore, the colony depended on imports, especially from Thailand, Myanmar, and Vietnam, and to a lesser extent, Java. These severe droughts did not just affect Malaya but were regional phenomena. Thus, harvest failures in those countries brutally impacted the availability and the price of rice across the region, a situation that in some cases lasted for up to two years, long after domestic drought conditions had subsided.[12]

The 1877 drought was to remain in inhabitants' minds for several decades. It was the same event that was cited as one argument for reserving forestland in Nathaniel Cantley's "Report on the Forests of the Straits Settlements in 1882, occurring as it did during the height of the nineteenth-century desiccation debate.[13] However, the severity of the drought also had the consequence of prompting government to do more to conserve and supply water as a more immediate and achievable goal. Although the government had an interest in the possibility of a connection between rainfall and deforestation, reserving the forests to preserve the economic resource was more prominent a consideration, and dealing with the town's immediate needs was more pressing than an esoteric and unproven scientific question. Indeed, while scientists and interested observers kept the desiccation question alive, by the 1930s much of the good conservation work that came from Cantley's report was being eroded in the name of development, especially on Singapore Island.[14] The colonial authorities understood the vagaries of the local climate in terms of needs-based responses rather than as an area of scientific investigation.

BUREAUCRACY AND THE PROBLEM OF MANAGING THE ENVIRONMENT

An anonymous contribution to the *Singapore Free Press and Mercantile Advertiser* in 1905 reveals the continuation of the debate over deforestation and rainfall that had become prominent in 1800s. The author referred to how Malaya's pineapple industry had contributed to forest clearances for both fruit growing and tin mining, the latter enabling Malaya to become one of the world's biggest exporters of the tinned produce. The piece also revealed the overlapping jurisdiction and conflict of interest between the Land Department and the Forestry Department, a classic expression of the problem of overbureaucratization within the Colonial Service.

> A generation ago we never had the temporary droughts that now occasion no small anxiety in connection with our rainfall catchment and distribution. We know that the Colonial Forest Department maintains what is a system of forest reserves, but these seem to be protected more for aboricultural than for hygrometrical reasons. . . . [T]he Land Department . . . is the one deadly enemy of our rainfall. For it is the only organisation whose ordinary working in the hunt after revenue is directed to the worsening of conditions on which the localized rainfall so greatly depends. The Forest Department and the Land Department are a house divided against itself.[15]

In the early twentieth century, the rise of scientific professions and professionals went hand in hand with the creation of new institutions and departments, which, while begetting more professional services, conterminously engendered ever more complex red tape and bureaucracy.[16] In Malaya, what this meant in practice was that a rash of new specialist departments took many of the joined-up responsibilities out of a few departments. The

Singapore Botanic Gardens is an excellent example of this splintering. In the late 1800s, the Botanic Gardens and its director had been in control of botany, economic produce, forestry, and agriculture for Malaya, but from 1901, this was all to change. As Barnard and Lee point out, the restructuring of such services derived from the emphasis placed on the "utility of science" for the colonial enterprise under Joseph Chamberlain as secretary of state for the colonies. The separation of responsibilities into a series of new departments was intended to facilitate greater specialization and more dedicated staff, resources, and expertise, but it has been argued that, although the separation did achieve these aims, it also led to taking the environment out of the hands of botanists and placing it into the hands of bureaucrats with all the associated problems that this entailed.[17]

The first area of expertise to be transformed was forestry. A Forest Department was created in 1901. It was headed by A. M. Burns-Murdoch, a forest officer who had seen several decades of service by this time in India and in Burma. Just as the anonymous author of the newspaper piece cited above suggests, this new department's raison d'être was resources management rather than environmental or climatic protection per se. Based on the Indian and Burmese models of reserved and unreserved forest areas, the emphasis was on protecting commercial forest produce, and although the new department had a duty to reforest areas that had been overexploited, this actually meant replanting with species of value to colonial interests not the reassertion of indigenous biodiversity.[18] The second field to move out from under the Botanical Gardens was agriculture in 1905, following pressure on the British Resident General and the British government by commercial interest groups in the region, including the powerful United Planters'Association of Malaya.[19] Indeed, in the first years of the newly established Department of Agriculture, there was little attempt to engage with smallholders or local farmers, instead their focus was on making connections with the plantation owners.[20]

It was during this new era of professionalization that the first Malayan Forestry Conference was inaugurated by the new Forestry Department in 1915. Their conference program was focused on consolidating the current state of knowledge and, from this, highlighting avenues for future investigation. Although their resultant agenda was delayed by the ongoing demands of the Great War in Europe (many officers left to enlist, and funding was rediverted), it was evident that there was much to be done. To this end, the latest director, G. E. S. Cubitt, appointed a dedicated forest research officer, F. W. Foxworthy, a well-known American botanist then working in the Philippines to come to Malaya.[21] One of Foxworthy's key priorities was to set up a dedicated research facility. He identified a large area of state-owned land at Kepong, just outside of Kuala Lumpur, that could be taken over by the Forestry Department for an experimental estate. The land was once virgin jungle but had been deforested over a few decades because of open-cast tin mining and had subsequently been used for vegetable farming, resulting in large swathes of eroded and infertile land.[22] After a swift and undoubtedly unpopular process of evicting the remaining farmers, the area was opened officially as the Forest Research Institute in 1929 under Foxworthy's leadership.

Just like the Forestry Department, though, the Forest Research Institute's research remained firmly entrenched within the realm of agroeconomics, with subjects including the extraction and sustainability of timber and forest products or experiments on the durability of wood for furniture, for example, featuring as core research outputs. Burns-Murdoch's last major publication before he died in 1914 is a fine example of this, the very first line of the preface stating how the text was intended to enable "Forest Officers and those interested in or dealing in timber to distinguish the most useful trees in the forests."[23] The first publications undertaken on Malayan themes by Foxworthy in 1921 and 1922 respectively, *The Commercial Woods of the Malay Peninsula* and *Minor Forest Products of the Malay Peninsula*, continued this theme.[24] As James Watson, who succeeded Foxworthy, noted in 1928, although the Arboretum reforestation project at Kepong was successful, the trees that eventually reforested the area were considered as much an experiment in growing valuable species and a form of "advertisement and propaganda, for with it we can demonstrate our ability to grow trees at a price . . . generally speaking . . . commercial trees will take precedence if space is at all cramped."[25]

Going back to the newspaper item cited above, in all areas of environmental research, nascent environmentalism or conservation was motivated or superseded by the desire for commercial success, and by the standards of the day this was not necessarily considered detrimental. Thus, the conversation about reserving forests to protect the climate paled into insignificance in the face of monetary matters. However, the problem of drought remained, and regardless of how people thought it might be caused (through man-made or natural processes), it needed to be addressed, as drought could periodically threaten the health of the agricultural and industrial economy. Thus, the government's preference in managing water security was to control and reserve extant water supplies as an immediate response to the weather, over and above directing expertise toward enriching scientific understanding of periodic drought. It may be, however, that the perceived lack of risk in Malayan society played a part in the government's attitude toward water security.

EXPERIENCING AND RESPONDING TO DROUGHT

The lack of risk in the tropical colony was summed up by Charles Burton Buckley in 1902 who had argued how "there are volcanic eruptions in Java . . .

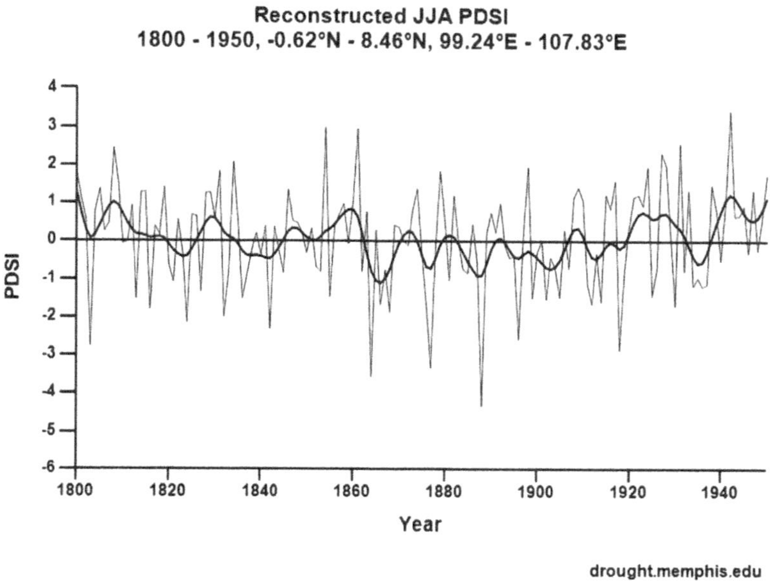

FIG. 4. Reconstructed Palmer Drought Severity Index (PDSI) for 1800–1940, using data from the Monsoon Asia Drought Atlas (MADA). Reconstructed JJA PDSI 1800–1950 created from the MADA online tool: http://drought.memphis.edu/MADA/Default.aspx. See E. R. Cook, K. J. Anchukaitis, B. M. Buckley, R. D. D'Arrigo, G. C. Jacoby, and W. E. Wright, "Asian Monsoon Failure and Megadrought during the Last Millennium," *Science* 328 (2010): 486–89.

earthquakes in Manila . . . typhoons in Hong Kong . . . cyclones in Calcutta . . . while, in Singapore, convulsions of nature are unknown."[26] It is important to reiterate that the sense of the tropical climate of the peninsula as one where "time ceases to be reckoned" by change of season and "all is constant midsummer" is one intrinsic reason behind the absence of risk management.[27] Indeed, the sense of stability that people felt may be borne out in modern precipitation data and evidence for early period nineteenth century. The reconstructed Palmer Drought Severity Index (PDSI), for instance, suggests some of the most dramatic dry events occurred only in the second half of the nineteenth century, with the first very severe drought episodes within that period around 1864.

The above dataset concords with narrative evidence compiled from historical archival documents for the same period, where dry periods are mentioned approximately every ten years or so until 1864 when a significant event appears. The dry periods that feature on the PDSI reconstruction above for the first half of the nineteenth century feature only minimally in contextual narrative, but they do correlate. The narrative sees mention of minor events in 1823, 1832, 1842, 1849, and 1855, and very little for other years. This may simply be because of the absence of records for those early decades, but the coincidence of the PDSI and narrative is striking. The PDSI also shows more severe events from 1864 onward, and this is reflected in a higher number of narrative records that mention a severe drought for each of the years 1864, 1877, 1885–1886, and 1894–1895. Similar patterns are also apparent in historical climatological reconstructions of the earlier period for Vietnam and Thailand, which suggest a fairly wet period between 1710 and 1820 in Southeast Asia, with some potentially cooler decades during the Dalton Minimum, which created positive conditions for agriculture.[28] Thus there may be good reason why colonial Europeans only had memories of relatively stable tropical weathers in the first decades of British settlement, creating a popular image that, regardless of the increasing instability, may have endured into the next century through the handed-down memories and lived experiences of some of the first- and second-generation inhabitants, despite the better meteorological records available during the last decades of the century.

Indeed, during the first decades of British settlement, reports of dry spells in the press, for example, were frequently accompanied by turns of phrase such as "in the memory of the oldest inhabitant" and tended to be noted with some surprise. Commenting on an extraordinary four-month dry spell that had affected the nutmeg crops in Province Wellesley in 1834, for example, one Dr. Lumsdaine had noted how "a drought of such continuance would not be thought extraordinary in many tropical regions; but here where there are frequent showers and heavy dews nearly throughout the year the unmitigated fervour of the Sun's rays was felt severely."[29] In Singapore in April 1837, a report on the weather remarked how even "the oldest European resident in the Settlement does not recollect any year the first quarter of which can be compared with the last three months, so very unusual has the season been. Instead of heavy and continued rains which might have been expected . . . the drought has been unprecedented even for our least rainy months."[30] The odd report of difference began to creep in, such as that of Penang Island by James Low, EEIC officer and frequent contributor to scholarly journals on the climate and environment of the region, who had observed that severe droughts were not random but "occurred at intervals of every four or five years."[31] Notwithstanding, even with the benefit of longer experience, reports of droughts still inspired astonishment in the late nineteenth century. In 1875, for example, there was a drought of such severity that "the oldest inhabitant, we believe, has no recollection of," and in 1899 "the oldest inhabitants" of George Town "sa[id] that wells are dry which, within their memory, were never dry before."[32] Given that the colonial memory was considered above that of other forms of remembering and reporting, and it was at that time limited to no more than a generation or two, it is perhaps one reason that the local political will to facilitate better hydraulic management failed to mitigate water shortages until later in the settlement's history.

The lack of political will to improve hydraulic management was also intimately connected to resources, money, and a lack of cogent planning. Under nineteenth-century British administration, water supplies in Singapore and George Town were managed by two separate agencies: the rural board and the municipal council. As we saw in the case of flooding, the lack of communication between the two could result in problems.[33] It was municipal council officials who were responsible for water in the early settlements, but aside from the basic provision of wells and storage tanks (predominantly intended for providing fresh water for ships docked in port), there was little provision for reservoirs and water infrastructures.[34] As the towns grew fast, they were therefore troubled by supply inadequacies and a hydraulic infrastructure that did not keep pace with the needs of trade, industry, and population. Although each town had a long history as a port, the arrival of the British changed the nature and scale of both regional and global commerce through the imposition of a free trade system and the inclusion of those ports within Britain's global trading network.[35] With the concurrent development of the plantation economy, the regional trade in spices, and the scaling up of the tin industry, both ports—especially Singapore—were catalyzed into major nodes in an imperial commercial trading network. Unsurprisingly, people followed quickly behind, including many who left their home country in search of a better life and opportunities.[36] Maintaining a busy fruitful port, therefore, placed exceptional demands on water. For both port towns, although there was a clear need to improve water reserves and supplies in the face of increased need and regular drought, the pace of change was slow and usually reactionary during the nineteenth century, rather than coming about as result of a coherent, advanced, and planned town policy.

VULNERABILITIES AND IMPROVEMENTS

In the field of disaster studies, it is commonly accepted that the societal structures that were in place at the time of a shock are just as important, if not more so, in turning a hazard into a disaster. While disaster studies themselves are largely a post-1950s phenomenon, governments have been dealing with disasters forever, as part of the role and expectation of governors in any society, culture, or period. Until the 1950s, however, and the rise of different methodological approaches to understanding the causes of disasters, historical governmental responses were often more reactive than proactive, and the effects of meteorological droughts on contemporary societies were largely dependent on socioenvironmental preconditions.[37]

Certain preconditions tend to lead to worse disaster outcomes. In Malaya, the period 1880 to the 1940 had the steepest population rise in urban areas since any of the settlements had been established, and there was immense pressure on land and housing as well as water resources. From the first census, taken in 1824, to that taken in 1871, the population of Singapore rose on average about ten thousand people per decade. From 1871 until 1947, the population increased on average by around one hundred thousand per decade, with the sharpest rise taking place across the 1920s and 1930s.[38] On Penang Island, for example, the population by the late 1920s was a little over 176,000, suggesting a rise of around 10,000 that decade alone.[39] The population increase was mainly catalyzed by immigration, over and above birth rate. In Singapore, the bulk of this migrant population was comprised of skilled and unskilled laborers many of whom were in the country on indentured contracts for fixed terms, and some were refugees from mainland China.[40] Many of these migrants came to the country poor and resided in the colony on a credit system, meaning that they owed their passage and associated expenses to a credit broker and had limited money or possessions of their own.[41] This predetermined that large sections of the population remained living on the poverty line, without sufficient means to sustain themselves through a crisis.

As James F. Warren has noted, the resources most immediately available to many of these migrants were "their own physical strength and experience, money lenders, pawn shops, and the private charity of friends and voluntary associations," which may not have been accessible to them or sufficient for them to manage during a disaster.[42] It is clear from the literature of risk and vulnerability, societies are at their most vulnerable when they have limited or unstable incomes and a lack of family or governmental support.[43] As many of the mostly male workers would have lacked family support networks locally, they were reliant on developing strong extrafamilial kinship bonds from allegiances of place and language, even down to regional and dialectical difference, lending "a reservoir of energy, sympathy and misery" to these newly constituted communities.[44] There were also the secret societies, many of which had originated in China and were constituted on historical, regional, and linguistic lines. These societies were initially tolerated by the British government as they were expected to deliver support to their own communities and appeared to provide some structure and organization within the Chinese population, a social group that the British rulers had little access to. Over the course of the nineteenth century things changed, however, as these societies appeared to be (in the eyes of the British government) at the root of various riots and crime syndicates and they were also accused of abuses within the labor markets that they controlled. In 1890 these societies were reconstituted as voluntary associations, after they were banned by the British government, a transition accompanied by a massive, deliberate destruction of society records.[45] Because of this event and the silence of the societies themselves during this era as to what they were or how they worked, it is hard to say to what extent they afforded their members assistance during times of crisis. It is known, however, that one overarching raison d'être of these organizations, especially after their reinvention in 1890, was to give mutual aid and charity to members during periods of crisis.[46] The high suicide rate of Chinese rickshaw pullers during the period 1918 to 1939 that Warren uncovered—particularly during the economic depression of the later 1920s—perhaps suggests that even these charitable institutions and extended kinship networks were not always able to alleviate the suffering and endemic poverty for many in this particular social group.[47]

It was also these poorer inhabitants of Malaya who were most vulnerable to the fluctuations in food prices that resulted from regional droughts. Take the El Niño–induced drought in 1911, for example. Domestically, in rice-producing districts such as Province Wellesley, the drought created an estimated 40 percent drop in rice output, which caused inflation to hit the remaining stock. Most of Malaya's rice, however, was imported from Thailand, where the drought caused a near failure of that country's harvest. The resultant rice scarcity meant that the British government had to shift its market suppliers to other areas, including Burma, which entailed a huge increase in prices. With the cost of rice in 1911 double that of 1910 and no increase in wages, it was noted, the "coolie classes" were struggling to manage.[48] Somewhat ironically, many urban dwellers were simultaneously subject to increased water rates, levied by the municipal council to cover improvements to the water supply, a move that affected all inhabitants living in the central areas, wealthy or not.[49]

This round of improvements had begun in the early 1900s. By now, Singapore had three small reservoirs in addition to the Impounding Reservoir, including one at Mount Emily to the east of the town center and one at Pearl's Hill to the west, a piped water supply connecting civic buildings and some domestic residences and there was investment in urban sewerage systems.[50] Primarily intended as a means to improve public health, plans for sewerage systems were designed by Singapore's municipal engineer Robert Pierce, and they were adopted in 1911 to provide a network of sewers, a pumping station, and a treatment and disposal works.[51] This network was connected to a new reservoir that had been created by building an embankment across the Kallang River valley, which was later named the Pierce Reservoir, in honor of its designer.[52] Like many of the government's schemes, however, the network took decades to implement and did not replace existing provisions for sewerage disposal but only acted in conjunction with traditional forms of manual night-soil collection. Despite the system's being upgraded and expanded once more in the early 1940s, only a small proportion of the town's buildings were ever linked to the central sewerage system until far into the postindependence era; the night soil collector was a familiar sight on Singapore's streets even in the 1980s.[53] The system also ironically contributed to pressure on water supply because of the increase in water consumption and wastage via flushing toilets, overrunning water, and leaky taps, sinks, and storage systems.[54]

In the 1920s a former president of the municipal commission, F. J. Halifax, explicitly pointed out how the improvements to supply undertaken since the early 1900s, including the Kallang River impounding reservoir completed in 1911, still did not provide enough filtered water for residents. For at least three or four months in every year, he stated, unfiltered water—which tended to cause deposits in the pipes but was mostly healthy—had to be used as substitute, though he did note that the two reservoirs (Kallang and MacRitchie) were able to sustain the town for several months during a water shortage.[55] The latter point was surely an exaggeration as evidence from the contemporary press reveals that water was restricted to only a few hours per day at the height of the 1911 drought. Other reports suggest that, in March that year, "coolies" were leaving Singapore in their droves because of the shortage of water and those that remained were engaged in frequent street fights over the standpipes, including one event that purportedly involved two rival *kongsi* (Chinese clan associations), about four hundred men in total.[56] Three months later, an open letter bemoaning the terrible insanitary state of the town because of the drought noted how "many Chinese who had the misfortune to live in the vicinity of these plague spots were frequently seen sweeping the garbage up into heaps on the roads and setting fire to it."[57] Tests were done on the quality of the water supply, and it was found that although the water quality was not dangerous, it had certainly reduced in quality owing to the drought conditions, and fish sold at the market was found to be rotten because of the shortage of ice.[58]

The discussion concerning urban water supplies was closely connected to the narrative of improvement, especially after the close of the First World War. Nationalistic and anticolonial sentiment often coalesced on the issue of social and health reforms, pushing nineteenth-century British paternalism into a new phase that advocated enhancing standards of living and well-being in the colonies.[59] Mortality, disease, and living conditions frequently related to water, either because of polluted supplies, accessibility, or simply not enough water. Percy Gerrard's 1913 essay on hygienic labor management, for instance, made explicit reference to the importance of a plentiful clean water supply for "coolie lines" such as there were in the western part of Singapore. Gerrard argued for wells and also the conveyance of pure water from nearby streams via a system of bamboo aqueducts, even the collection of rainwater if these supplies proved insufficient.[60] His remarks come as something of a surprise as, already by the mid-nineteenth century, wells were frequently associated with contaminated water.[61] During the drought of 1877, for example, dried-up wells providing brackish water were the source of a cholera outbreak purportedly resulting in the deaths of many people within the Malay community at "Guthrie's Land," near Tanjong Pagar in Singapore.[62] In 1902, another severe drought led to speculation that old disused wells ought to be reopened to alleviate the contemporary water shortage, a claim refuted by Dr. W. F. C. Middleton, health officer for Singapore, on the basis that they were closed for the very good reason of being polluted.[63]

Middleton had been a key figure arguing for the closure of such wells ever since the 1890s, contending that their use was one of the reasons for the town's high mortality rate as they operated as breeding grounds for diseases including cholera.[64] To put this into perspective, the Singapore municipality in 1901 commissioned a survey of remaining wells within town limits, and it was discovered that out of 3,877 existing wells, only 609 contained potable water.[65] The headache for the health officers was that, during times of drought, people tended to take matters into their own hands. During the 1902 drought, for instance, when the municipal water supply was restricted to only two hours per day, a number of bricked-up wells were reopened by urban inhabitants trying to improve their lot—plausibly those who were not able to replenish their supply in the one-hour-twice-per-day time slot allocated.[66] The bricking up of wells may have been done in the spirit of public health but was not universally popular. The limited supply, only available to the poorer inhabitants at street-based standpipes, led to a temporary black market emerging for the sale of water.[67] The result was that the municipality had cracked down on what they saw as "crime," arresting malcontents involved in fighting over the limited resource and issuing fines to people who were discovered to have reopened polluted wells.[68]

Well water was regularly tested for contaminants by the local Health Department, as were the town's water supplies from the late nineteenth century. Extreme rainfall—whether drought or flood—was known to be a risk for contamination. During the 1911 drought, for example, samples from ten municipal wells were analysed to check on the quality; although no life-threatening pollutants were identified, it was discovered that the water quality had deteriorated.[69] Testing became far more sophisticated in the 1920s with the municipality employing a bacteriologist along with an analyst whose job was to test well water and municipal water supplies, alongside various types of effluents, sewerage, and foodstuffs. Well water in general, drought or no, always came out with higher levels of pollutants, which, according to municipal analyst A. G. Harrington, was due to their age, poor construction, and proximity to urban housing, which left them liable to seepage from domestic sewerage. In his opinion, all the town's wells ought to be closed.[70] James F. Warren's study of the social conditions of Singapore's "rickshaw coolies" pointed to similar conclusions, arguing that the water in the urban environment was hazardous in both quality and quantity, contributing to a high mortality rate in this largely impoverished group of urban inhabitants, with intestinal and diarrheal disease high on the list of causes.[71] While there was a concerted effort to follow Middleton's advice and get all the wells finally closed during the 1920s, this had the unfortunate consequence of pushing the laboring classes to use municipal standpipes, which were not as convenient, and during drought periods, they became places of contention.

ENGINEERING MODERNITY

The triumphalist narrative of modernity that underpinned colonial hydrological projects often belied their social consequences.[72] The continued dependence of urban dwellers on sources of substandard water, despite the existence of technologically innovative solutions, was not only a problem in terms of health but also pointed to a wider cycle of poverty and exclusion from the public sources provided by the colonial government. The government in turn presented solutions that furthered dependency on bigger, grander, more complex, and more expensive technocratic projects, resulting in a situation similar to that which Mark Elvin termed "technological lock-in" in the context of hydraulic schemes in imperial China.[73] Ever more reliant on scientific engineering and the advice and buy-in of experts, grand hydraulic schemes distanced the control of water from ordinary urban dwellers, through complicated and expensive means of access and by turning water into a scientific and state-controlled problem mediated through the role of the expert.[74] The rise of scientific experts was closely connected to the emergence of the state—or, in the case of a colony, the rule of the colonial power—and was closely aligned with social engineering in the name of progress and modernity.[75] By way of example, the scale and technological reach of proposed means to ensure an upgraded water supply became more and more extravagant into the twentieth century. The premise underpinning the most ambitious plan rested on the fact that there simply was not enough of a usable water supply within Singapore itself.

As early as 1904, Pierce was wondering whether the rural areas of peninsula Malaya with their abundant rivers, streams, and springs might be tapped to ameliorate the shortfall in town. Having undertaken research on the Kallang water scheme and another at Seletar within Singapore's island borders, he was convinced that these would only be short-term solutions. However, the sheer scale of transporting water across the Johor Strait seemed beyond the technological (and financial) imagination of the municipal council at that time, and the Kallang Reservoir plan, agreed upon and started in 1902, was still expected to be of great value to the town, promising a daily supply of nine million gallons of water. By the time that the scheme had opened however, the predictions of water usage had already far exceeded the reservoir's capacity. There were also other problems inherent in the building of new urban infrastructures for water, because of the unique nature of Singapore's environment:

> One of the difficulties of reservoir construction in Singapore lies in the fact that the country possesses no high hills and no steep valleys. Any reservoir therefore constructed as this one is by damming up the course of a river, is of necessity shallow in depth and large in superficial area. This reservoir for instance is only 27 feet in depth but has a water area of 253 acres. It is easy to understand that the exposure of so large an area of water to the rays of the tropical sun must mean a large loss of evaporation . . . about 6 inches per month . . . which represents a loss of no less than 400,000,000 gallons of water each year. A further disadvantage of the exposure . . . is that the temperature of the water is increased and the growth of microscopic forms of both animal and vegetable life is encouraged, necessitating expensive filtration works.[76]

In 1902, a major feat of colonial engineering was proposed: the Johor-Singapore Causeway. Ostensibly built to facilitate improved land transport for goods between mainland Malaya and the port of Singapore, the rubble-based causeway would allow rail and road transport but was also intended to support a pipeline for water.[77] The contract was awarded in 1919, and the first trains began to run in 1923.[78] Various tenders and plans for the conterminous water project were considered, but it was only in 1922 that a scheme proposed by consulting engineers Alexander Binnie, Sons and Deacon was accepted, subject to approval by the government of Johore, for the supply of water from Gunong Pulai to Singapore town, which would then be stored at reservoirs at Fort Canning. The scheme not only involved the building of the reservoir at Gunong Pulai but also significant work at Fort Canning, the end destination of the piped water, as well as the infrastructure required to transport the supply across the thirty miles from site to site.[79]

The scheme was described as one of the most ambitious departmental enterprises in Britain's eastern empire, requiring a network of pumping stations and electrical stations, workshops, foundries where machinery and apparatus were manufactured on site, laboratories for testing, and a multitude of skilled mechanical engineers.[80] This would replace original proposals to construct a small reservoir at Johore in conjunction with one at Seletar within Singapore, although it was proposed to undertake waterworks at the Seletar catchment nonetheless as an interim measure. The immense costs, estimated at $30,000,000 (but in actuality they came to far more), were expected to be mitigated by rates charged for the water itself, with capital costs recovered from a Sinking Fund over a period of from forty to sixty years.[81] Like the Impounding Reservoir (MacRitchie Reservoir) of the previous century, the Gunong Pulai Reservoir operated on the gravitation principle and was opened officially by the sultan of Johor on 10 April 1927. The pipeline opened a few months later, and the whole scheme was extended in 1935 by an additional reservoir at Pontian Kechil, which was linked by a series of dams to Gunong Pulai on the Malayan mainland.[82]

Delivering the additional water supply was originally considered "so necessary and so urgent," however, "that every facility will be given for the most rapid construction of the works."[83] The scale of the need was exemplified by engineer Colonel Pearson, who modeled the colony's projected daily usage of water in conjunction with population growth.[84] Noting how his modeling was simplistic and did not account for other factors such as the weather, which could tip the balance one way or the other, he cautioned that any delay in starting the works would risk having to live through another drought without an adequate supply, which would result in a reduction of the immediate environmental quality of the town from dust pollution, fire risk, limitations to important industries such as rubber production, and reduced hours of supply for public consumption and for hospitals.[85]

The job of predicting need was complex. It was based on estimates of intangible factors such as hygiene (bathing and washing clothes), cooking, and cleaning as well as drinking—alongside the larger-scale more overtly tangible needs of trade and industry. It has been suggested that until the wide-scale introduction of water meterage and other means of scientifically tracking consumption any estimates were simply that, fashioned by scientists and engineers articulating present and future, through modeling but also sociocultural assumptions and political requirements.[86] In 1938 David Murnane, then Singapore's municipal engineer and a significant figure in Singapore's water history, noted how the job of supplying water was an exercise in seeing into the future by understanding the past, generating models that accounted for variables such as evaporation and rainfall. Long-term planning, in his words, meant the "water-works engineer spends his time carrying out the plans of his predecessor and thinking out the work to be done by his successor."[87] It was Murnane who oversaw the completion of the project begun under Pearson. Taking over as a water engineer in 1925 and playing an instrumental role in postwar reconstruction until his retirement in 1947, Murnane survived the Second World War in one of Singapore's many civilian prisoner of war camps.[88] With a heavy dose of irony, he witnessed the dramatic closure of the very supply that he had been a part of delivering.

When Japanese forces entered Malaya from the border with Thailand in early December 1941, they made a rapid southward progress down the peninsula attacking strategic sites on their way.[89] On 11 December, they reached Penang, which fell quickly to the invading forces after heavy bombing resulted in a significant loss of life and damage to urban infrastructure including water and electricity supplies.[90] The retreating Malayan forces and their Australian allies operated a scorched earth policy, adding to the destruction of key infrastructure, including electricity and water supplies, airfields, and defensive positions. By the end of January, the capture of the peninsula was almost complete, and the Japanese military began their assault on Singapore on 8 February 1942. The Johor-Singapore Causeway was the only land crossing between mainland Malaya and Singapore. In a combination of Japanese shelling and hasty British strategy on retreat, the causeway was partially destroyed in January 1942, severing the largest supply of water into the town.[91]

The causeway's design had been contentious in some naval circles as it was an earthed structure without the capacity for a cantilever bridge allowing for passage of vessels in that area.[92] What had also not been considered at that time was the possibility of an invasion by land from Malaya, and a different design might have alleviated the need for the causeway to be partially destroyed and thus cutting off the water supply. For the troops and people left in Singapore, the obliteration of this supply signaled a turning point in the war. With the main supply to the island cut, British Army Commander General A. E. Percival reported that Singapore would run out of water imminently. The island's supply at MacRitchie Reservoir along with the service reservoirs at Pearl's Hill and Fort Canning had been damaged, leading to a serious loss of water pressure, although the supply remained intact, and the Pierce Reservoir had been polluted by dead bodies following an attack.[93] These combined factors were the straw that broke the camel's back. Historians have subsequently argued that the lack of water was not as critical as was imagined at the time, but the fact remains that the perception of imminent failure was the catalyst that led to the British surrender on 15 January 1942.[94]

THE END OF DROUGHT IN URBAN MALAYA?

The Gunong Pulai scheme had been vaunted in 1926 as the end of drought for Singapore.[95] In January 1930, a contributor to the *Straits Times* noted how the engineering feat was a clear success; despite the failure of the northeast monsoon that December, "there is no need for anxiety . . . the water in the Gunong Pulai reservoir is within a foot of the top of the dam, which is now holding up something over 1,200,000,000 gallons of water."[96] That article was to herald a significant drought that lasted for much of 1930 but it seems that the writer was correct: there was little immediate impact on the town.[97] Indeed, this was to be the case in the urban areas until the outbreak of war in 1942, with droughts registered in the press and in meteorological records, but their mention was not accompanied by the furore attached to earlier events.

Nonetheless, this reading would be too simplistic. By 1939 the water supplies to urban areas—although far better than a generation earlier and now with piped, filtered, and metered water extended beyond the municipal borders—were still not sufficient for need given the continued increase in demand from population, industry needs, and the introduction of the waterborne sewerage system.[98] This was abundantly clear to the engineers who renewed their call to extend the Johore–Singapore pipelines and to create a permanent reservoir at Seletar.[99] In October 1939, A. N. Galsworthy brought a plan back to the attention of the legislative council and the treasury, which had been put aside because of the outbreak of the First World War. The engineers were arguing that the daily consumption of Singapore was twenty-three million gallons, when the output of the water supply was a million less; the supply neither satisfied the town's needs nor left any buffer for dry weather.[100] In light of the circumstances created by the European war, however, there were problems securing the loan of five million Straits dollars deemed necessary to begin the work. Treasury officials had to argue hard to secure the loan, and although they received it finally, the terms were altered to become less financially favorable to Singapore.[101]

And, in 1939, even while Singapore and George Town had seen extensions and improvements to their supply, the rest of Malaya and even outlying parts of Singapore Island and Penang Island were little better off than they had been years earlier.[102] Lacking a proper linkage to the water supply, residents of the southwestern village at Tanjong Kling in western Singapore, for example, had undergone "much hardship" during a drought in early 1940.[103] This drought, result of the failure of the 1939–1940 northeast monsoon, was considered very serious across most of Malaya and it did not end until well into the autumn of 1940. In Kuala Lumpur, it was considered one of the worst droughts for many years, although there was no outright supply failure, despite significant water shortages at its main Kuala Pilah reservoir.[104] Smaller towns fared worse. In Kuala Kangsar (north of Ipoh, a mining district in Perak), the water supplies were cut during the daytime, and surrounding estates and villages found that their wells had dried up.[105] It was the smaller villages and rural areas that really suffered.

In the Malacca district, one of the hardest hit areas, the drought was considered the worst "within the memory of the oldest inhabitants. 'Never before in the history of the Settlement have I heard of our streams drying up,' said a *penghulu* [headman]. . . . Cultivation of any sort is impossible."[106] The drought had threatened the *padi* (rice) harvest; fields that had been ploughed ready for planting were "as hard as rock"; fruit, including local staples such as rambutan and durian were smaller or even dying before they ripened.[107] Across the peninsula, the supply of water to rubber estates was limited, affecting yields for estate owners large and small and severely impacting on the quality of life of the estate workers. The Public Works Department drafted in trucks to provide water to some estates from better supplied urban areas, as well as to some of the villages where wells had dried up.[108] In the Alor Gajah district, just north of Malacca, residents petitioned Governor Shenton Thomas for urgent assistance as their water supplies were cut back to sixteen gallons per household per day, delivered in old kerosene cans.[109] British Resident J. V. Cowgill at Negri Sembilan, the district between Kuala Lumpur and Malacca, argued that the government had to do something to avoid a repeat of the conditions experienced there during 1940, when water had to be carted from nearby towns into villages just for people to have enough water to drink.[110]

Rural shortages also created a feedback loop for urban inhabitants' resilience to drought. Because of the severity of the water shortages in the Malacca district in 1940, for example, the padi (paddy) harvests in that area failed. The government was obliged to provide $63,200 in relief works and a free distribution of padi seeds intended to rehabilitate the crops. But there would have been a knock-on effect in terms of local food prices and availability in the short-to-medium term until new harvests could be sown and reaped.[111] It was commonplace, too, for the many Chinese communities inhabiting peri-urban areas to grow vegetables for nearby town markets across the region, including in Singapore and Malacca. In the 1940 drought, these vegetable suppliers were hit severely, which set off a chain of events impacting the price and quality of surviving fresh vegetables for the town dwellers.[112]

This coincided with concerns about food security more generally, which had been mounting across the 1930s and reaching new levels since the outbreak of war in Europe in 1939. In an unfortunate turn of events, the drought led to the postponement of a new raft of agricultural schemes designed to increase food productivity and improve self-sufficiency in the event of further wartime trade disruption. The scheme—vaunted by the new Food Supply Office at least two years earlier—centered on devoting more space to and promoting education on urban and peri-urban gardening to produce vegetables and fruit, instead of relying on imports.[113]

Thus, not only was there a distinct urban-rural divide in the availability of water but the reliance of the towns on the rural areas for fresh food meant that, regardless of improvements to the urban water supply, there was still a point of vulnerability as it related to food, especially for those who could afford less. At the outbreak of the Second World War, there was still a divide between rich and poor and between urban and peri-urban, with many inhabitants still relying on wells and standpipes as they always had done. As Singapore resident Ang Seah San recalled of his time in occupied Malaya between 1942 and 1945, although the urban water supply was never too much of a problem, in the countryside "most of them were dependent on the wells for their water."[114]

When the British government regained control over Singapore and Malaya in September 1945, the interim British Military Administration had the task of assessing the damages and reconstruction needed for essential infrastructures, including water supply. During the war itself, the Malayan Planning Unit (MPU) was established under the auspices of the War Office. With next to no idea of what had been taking place in Malaya during the Occupation, the MPU was asked to put together a reentry strategy that relied on prewar records and the combined knowledge of the officers who were mainly ex–Malayan civil servants.[115] Oswald Gilmour was one of these men and one of the first back into Singapore after the Japanese surrender. Gilmour was put in charge of utilities and works, so within a few days of arriving back in the town, he was out at Gunong Pulai inspecting the waterworks. David Murnane was with Gilmour on this visit, recently released from the Sime Road civilian camp, and Gilmour reveals how Murnane was welcomed with open arms by the staff who had continued to work at the reservoir over the past three years in very difficult conditions and with little knowledge of what was happening at the other end of the pipeline.[116]

Singapore's water supply was in critical condition. It was still functioning, but with minimal repair work and maintenance over the three years of occupation the equipment had degraded. Many of the pumps were inoperative and some of the filtration plants had broken down, although the supply from Gunong Pulai still reached the reservoirs. In fact, in lieu of the pumps, the supply from Johore had been overused to the point that the reservoir was almost completely dried up and the two service reservoirs at Fort Canning and Pearl's Hill were empty. The municipal water meters were ruined, and of the water supply that remained much was going to waste through leaking and broken pipes. Added to this was the sudden influx of civilian and military population who brought the supply to breaking point.[117] This was also the situation in Kuala Lumpur, where "the water supply was untreated and thousands of gallons were running to waste from defective and tapless pipes."[118] Gilmour further recalled that by the time the civilian government returned in April 1946, countless problems remained, although Major A. C. Wilson, who had been placed in charge of the Water Works Section, had the basics under control. By 1947, Murnane was warning of potential water rationing to be brought in over the next two years, not because of drought but simply because supply would be unable to match demand—the new works that had been started were far from being finished.[119] Despite more than a century of water management, in 1947, postwar Singapore's reservoirs were struggling to meet day-to-day needs.

THE MIRAGE OF WEATHER CONTROL

In colonial Malaya, the emphasis was on controlling water shortages through reactive engineered solutions rather than through investing in forward planning and storage that would keep up with projected population, usage, and frequency. Neither were new technologies considered to combat drought, such as cloud seeding, despite their emergence in countries across the world including the United States, Russia, China, and other British colonies such as India and Hong Kong during the 1920s, 1930s, and 1940s (albeit without much success).[120] This was very much in line with the colonial government's attitude toward flood control. In both scenarios, mitigation relied on providing ample engineering works to dampen down the effects of natural climatic undulations so as to create a new normative relationship between man and water. Yet the local authorities appeared to be always two steps behind. Engineering works replaced indigenous methods of living with the weather, with the expert and the technocrat replacing individual or community resilience but without managing to provide a satisfactory new relationship for everyone.[121]

There could have been more time and effort placed into understanding the average climate patterns of Southeast Asia, especially the action of the monsoon, and connecting this with engineering needs. Instead, engineers did little more than build annual averages into water usage models, as they had already been doing for decades. Even the probable action of ENSO was not really taken on board despite new awareness of this phenomenon in worldwide scientific circles (sunspots and deforestation were often still given as reasons for periodic droughts in the twentieth century), and the discovery of other coupled ocean and atmospheric phenomena such as the Indian Ocean Dipole was still far in the future. The discussion of cyclical weather patterns led to little direct action in Malayan circles, despite its being as a possible explanation for the regularity of drought periods.[122] Moreover, there was little investment in forecasting droughts or trying to determine their frequency beyond some basic observational notes, unlike meteorological experts in India or Australia had been doing since the 1870s.[123]

In the late nineteenth century, Henry Blanford's discussions of rainfall cycles, although for the most part focused on India and Australia, also spoke about Singapore.[124] Using meteorological registers of pressure from different parts of India, Batavia, Port Blair, and Singapore, he confirmed direct evidence of teleconnected observations and periodicity across the 1860s and 1870s, but it appears this was not translated across into scientific research being undertaken in Malaya even in the twentieth century. Perhaps the lack of dialogue from within Malaya or about Malaya was because droughts did not pose the type of long-term threat to productivity and settlement that was apparent for Australia or the slow disaster of famine as in India, thus there was little reason to undertake such research.[125] As Norman J. Lockyer wryly noted in 1872, the task of looking for patterns in weather observations was largely one of groping in the dark.[126] But perhaps this accusation is unfair and delivered with the wonderful application of hindsight; even in India, the dialogue on periodicity remained centered around Blanford's research for the rest of the century and beyond. In 1890, the discussion of correlation of high pressure, sunspot activity, and drought had once again raised its head but in direct relation to Blanford's now rather outdated observations and comments on the years from 1876 to 1881. John Eliot, Blanford's successor, noted how the features of excessive pressure from late 1888 to 1889 bore a marked similarity to those marked by Blanford in 1876 to 1878, including the correlation with high temperatures and minimum sunspot activity.[127] Maybe there was just a lack of scientists such as Blanford or Eliot to champion the cause in Malaya, with only a few trained weather experts in the country until after the advent of the MMS.

Even then, as vast improvements in climate and weather knowledge due to the technological advances of the era, especially aviation and radar, enabled studies of the upper air, Malaya still fell behind in how officials thought about drought or climatic oscillation. The state of research is encapsulated in the 1935 report by C. D. Stewart, superintendent of the MMS. In his commentary on contemporary weather forecasting in Malaya, he argued how prognostication in the equatorial belt was a special problem and there was a lack of precedent or guidance. Moving into a summary of the methods then in use, he pointed out that these were "still largely experimental."[128] Current research on rainfall, he explained, was conducted on the basis of wind direction, largely because of the "lack of serious variation in other determinate meteorological elements." This information was then plotted on a synoptic chart used to determine the air stream, in conjunction with prior observations of wind direction, dew point, and rainfall amounts.[129] These studies were geared toward short-term weather forecasting, rather than understanding the potential for multiyear climatic events such as droughts or heavy rains.

Moving forward, while Southeast Asian meteorological flights commenced during the 1940s, and weather radar and an upper air observatory were brought to Malaya in the early 1950s, the MMS's generally poor finances did not allow research beyond the immediate need of forecasting for aviation and the military.[130] The Malayan Emergency of 1948–1960 meant that a return to peacetime meteorology was not on the cards for the MMS following the end of the Second World War, as forecasting for anticommunist operations took priority over other research.[131] In 1948, the director of the MMS, H. B. F. Moorhead, explained how difficult it was to provide long-term predictions for the frequent inquiries from civil departments such as agriculture, irrigation, engineering, or insurance because of their postwar staff shortages and the uptick of workload for the military. Interestingly, Moorhead also pointed to the lack of information on rainfall patterns because of the Japanese Occupation where scorched earth policies on both sides had resulted in the loss of almost four years' worth of records. In his words, this loss decreased "to a considerable extent the value of the preceding 30–40 years [of records], which period was becoming long enough to permit of the calculation of reliable means of temperature and rainfall."[132] Thus, perhaps, the emphasis on engineered solutions is more readily explained, rather than any two-pronged approach of prediction and technological mitigation.

The inability to fully to provide a water supply that not only mitigated but obliviated the effects of extreme weather would continue throughout British rule. In 1950, the British government commissioned a survey of water resources on Singapore island, undertaken by consulting engineer Bruce White of Wolfe Barry and Partners, in view of the possibility of a supply interruption from Johore, likely inspired by the events of 1941 when the pipelines were bombed. The lack of localized water resources, however, meant that reliance on Johore would continue, and that waterworks would continue to fail during droughts. In 1962–1963, for instance, drought struck the whole of Malaya, creating terrible supply restrictions in many towns and villages and affecting rice production across the peninsula.[133] British Pathé news footage of Singapore released in 1963 shows open congested drains, low water levels in canals and river, and people walking to collect water in cans from standpipes in the street.[134]

These colonial-era legacies continued postindependence in some places. In Malaysia, water shortages continue to be a regular occurrence, despite the availability of abundant local water resources, and in Singapore, the dependence on Johore for water continued and became a major bone of political contention when the two countries separated formally in 1965.[135] To put this into perspective, one of the first actions apparently undertaken by the newly inaugurated prime minister of Singapore Lee Kuan Yew in 1965 was to call the chief engineer of the Public Utilities Board to ask whether it was possible for Singapore to become water self-sufficient.[136] Only now and into the immediate future is this vision of Singaporean self-sufficiency becoming a reality, enabled by recent technological developments in desalination and water recycling, but in Malaysia, many areas still suffer annually from droughts in addition to the repeated flood disasters in rural areas.

CHAPTER 7

REGULATING HEAT, CONTROLLING URBAN AIRS

Although tropical medicine and the understanding of the vectors for disease developed significantly in the hundred years or so after the British first settled on the Malay peninsula, heat and humidity were always considered a major challenge to tropical living. Dr. William Simpson noted in 1907 that "the moist heat suits some constitutions while others become acclimatised, but with the remainder, that is with the majority, it is a struggle to resist its debilitating effect for any considerable time."[1] After the introduction of electricity, however, new and better ways of managing the heat became available to some inhabitants, especially the wealthy. Electric fans, refrigeration, and (later) air-conditioning offered new ways of thinking about, moving in, and inhabiting a tropical atmosphere. While piped water and flood mitigation enabled some urban inhabitants to enjoy a basic quality of life, these new technologies offered a tantalizing vista of thermal comfort and modern living. This was a small-scale form of weather control that fed directly into people's lives, bypassing atmospheric research and moving straight into a situation where residents could live out of the weather within indoor spaces.

The notion of atmospheric control as a technological process arrived in an era of scientific modernism and futurism. Science and technology were synonymous with progress and, in the case of Malaya and other colonies, were entangled within the imperial system. By the beginning of the twentieth century, technical innovations such as wireless technology, electrification, photography, mechanical computing, and much more ushered in an era of technological optimism whereby technology was considered the answer to many human problems and was transforming the ways in which overseas empires operated. In the unforgiving climates of many of these colonized regions, the concept of engineering and modifying climates was appealing.

Across the world, French, American, and British governments in particular hypothesized the possibilities of engineering the environment by creating artificial lakes and forests to enhance precipitation or by transmogrifying desert and tundra into agriculturally fertile zones.[2] Such grand schemes were not attempted in Malaya, but the basic principle of managing an indoor environment by removing humidity and lowering temperature was not such a different discussion from that of reserving a forest to cool and moisten the local atmosphere, nor was the principle far removed from the imperial government's interest in mitigating the tropical climates' extremes. At the micro-level, indoor climate control and cooling technologies held the promise of transforming everyday living in hot, humid weathers, improving the productivity of workforces, and thus enhancing the productivity of empire.

Looking specifically at urban indoor atmospheres and the introduction of technological cooling systems, this chapter concludes the final part of this book by considering heat and humidity as the final atmospheric challenge for settlers on the Malay peninsula to manage, but—just like the discussion of extreme rainfall, flood, or drought—managing heat and humidity also involved the imposition of Western ideas, in this case concerns about being cool and, in so doing, created fresh social aspirations that built on and exacerbated preexisting social divides. As Thomas P. Hughes pointed out several decades ago, there are deep connections between the social and the technical. In the case of the advent of air-conditioning and the larger technical networks upon which it rested (electrical infrastructures, knowledge transfer, urban transformation), the social and the technical were bound together in relationships of power and ingress. In Malaya, cooling technologies were as much social identifiers as technical objects, and although they may not have superseded more traditional ways of staying cool in the tropics until the modern day, their introduction tacked onto expectations of social success within a Western value system. Who had access to cooled air was intimately coupled with socioeconomic power and integrated into an ideological system that echoes across time and space.[3] Just like access to clean water and dry housing, cooled air is part of a wider socioecological system of urban resilience.

This chapter focuses solely on the mechanisms and social implications of controlling indoor weather.[4] It argues three interconnected points. First, and despite the setting of this book, that the drive toward controlling indoor thermal comfort was not specifically British, though it was Western in some respects; neither was it about colonial power in the respect of enforcing control through a show of technological superiority. Modern air-conditioning grew out of early experiments with refrigeration technologies undertaken in the United States, though the concept of cooling air was not specific to that region. Indeed, air cooling had been practiced in parts of Asia, including

China and Korea, for centuries. This tallies with recent arguments that, despite early postcolonial historiographical rhetoric on the strength of British technological prowess, by the early 1900s America was more progressive in many fields of technology, and even then, many of the ideas to come out of the United States were not indigenous but had been adapted from elsewhere and "invented" under the US flag.[5] In Malaya, air-conditioned environments were exclusive spaces, though, and the people who benefited most directly from them were often, though not exclusively, a colonial class, not based exclusively on race but a community that thrived on the practices of colonialism. These were the plantation owners and the merchant capitalists who had grown wealthy on the structures and conditions offered by the colonies. Cooling technologies were embedded within colonial-era socioeconomic structures and networks. Their advent and adoption presaged a new way of distinguishing between rich and poor, and cooling technology became a marker of wealth and better living standards. Technology has always been seen as a precursor of modernity, but importantly, one that obscures the agency and voice of subaltern actors.[6] In Malaya, poorer families were unlikely to have air-conditioning or even access to the growing number of air-conditioned spaces that dotted the towns, unless they had jobs that directly connected them to these places.

Second, although Malaya's modern technological climate control methods were largely imported from the West, perceptions of what owning cooling technologies meant were not necessarily Western. Some postcolonial narratives argue that the import of Western technology was yet another form of cultural indoctrination and economic exploitation that went unrecognized because of the purported benefits (usually from Western literature) of said technology to health or to other forms of physical comfort. In Britain's colonies, the supposed superiority of technology was imbibed along with British history and literature, and the English language was held up as the language of technology and development. As Goh Chor Boon has argued, however, Malaya failed to develop a significant technological culture (and thus did not industrialize) under British rule. The focus of Goh's claim is on the uptake of imported technologies, which, he argues, was minimal until the 1990s, a fact that he blames on a conservative, traditional business culture and lack of interest in technical educations.[7] More than this, the British Malayan school system did not prioritize technical education; tertiary education of any form was limited for non-Europeans. I argue that the uptake of new technologies had more to do with money or simply a purposeful continuance of traditional ways of living than with an aversion to technology.[8] It must be kept in mind that for many people living in early twentieth century Malaya, heat and humidity were not an evil to be overcome but a simple fact of life. Thus, in the ensuing discussion of the introduction of cooling technologies, whom they impacted and how, we must bear in mind the fact that this discussion comes from a Western lens through which value is ascribed to refrigeration or air-conditioning, or a modern lens whereby urban heat is increasing in cities. Thus, while this chapter explores climate control as a form of colonial-class sociotechnical imaginary, this point ought to be tempered by awareness of these lenses.[9]

Finally, it also suggests—as historians of technologies have been telling us for some time—that the rise of technological solutions for controlling heat did not mark a straightforward transition into modernity.[10] New cooling technologies did not replace more traditional methods of dealing with hot conditions but coexisted alongside them far beyond the focus of this study. Thus considering not only the types of spaces that became controlled environments but also when that happened says much about how colonial-era aspirations to modernity were restricted to time, space, and person.

TRADITIONALLY COOL

For much of the nineteenth and early twentieth centuries, managing heat and humidity meant architectural solutions and avoidance.[11] There were Malay diaspora and indigenous traditional building designs that existed in the region long before the colonists arrived and that were well adapted to rural and coastal tropical living. These homes were designed to be malleable in response to the weather, built from local wood and *attap* palm often atop piles or pillars to allow air to flow beneath and through them.[12] Well-to-do Malay houses might also sport wall panels of carved wooded grilles and gable-edged roofs that allowed for ventilation and internal air circulation.[13] The homes of the ultra wealthy Peranakan or Straits-Chinese families also utilized space and air flow. Tiled floors cooled the feet, doors were built with ventilation grilles, air vents punctuated the walls, and louvered shutters allowed air flow while retaining privacy and security. Some homes were built around a central air well, open to the sky and the rain, though mostly protected from the sun by the roof design.[14] The mansion that belonged to industrialist Cheong Fatt Tze in George Town had a most ingenious internal cooling system with pipes that captured rainwater from the roof and then channeled this through walls and floors to cool them.[15]

The domestic and civic buildings that sprang up as a direct result of colonization incorporated many such ideas to mitigate the extremes of heat. Hybridized from a combination of European (especially Mediterranean) design, indigenous knowledge, and the British experience in India, cooling strategies included thick walls to keep out the heat, verandahs covered in blinds that created air flow and provided a double layer to direct the sun away from a building's main walls, and occasionally an open space or "air well" was incorporated into the center of the building. Inspired by Indian and Caribbean designs, the *chick* (blinds) and *tatties* (rolled-up screens) were sometimes keep wet, ensuring that the incoming air was moist and cool.[16] This hybridization was important to contemporary architects such as T. Roger Smith, who argued that tropical buildings, while addressing "climatic disturbances . . . wind-storms, dust-storms, thunder-storms," heat, and humidity, ought also to appear as "English if possible; or, if not at any rate European . . . with such modifications as the climate, the materials, the power and training of the workmen, and other local circumstances render necessary."[17] This may well have been important to the European occupants of such buildings, too, who sought familiarity in their surroundings. In Singapore's civic district, successive architects and civil engineers—including the very first, George Drumgoole Coleman—adopted a similar approach in designing public and government buildings. Appointed Superintendent of Convicts and Public Works in 1833 (the seemingly incongruous roles worked well together as most of the early civic buildings were built using convict labor), he brought with him previous experience from Calcutta and Batavia. His building plans incorporated many of the features familiar to the British colonists, including the Neo-Palladian style, and the heat-repelling verandah and louvered window.[18]

Location was also important. The Istana Lama, home to Temenggong Abdul Rahman in 1824, was built on coastal high land at Telok Blangah (later Mount Faber), which would have been a prime location for catching refreshing sea breezes. Further inland but still on higher ground was Istana Tyersall, estate of Temenggong Abu Bakar of Johore at the western edge of the Singapore Botanic Gardens. The bungalows of wealthy residents —such as those at Tanglin Hill, Nassim Hill, and Mount Alma to the northwest of Singapore town near the new botanic gardens site and off the Orchard Road at Mount Elizabeth and Goodwood—were situated on higher land within ample gardens to generate maximum exposure to breezes and cooling vegetation. Eden Hall, for example, the mansion of a successful trader, Ezekiel Saleh Manesseh, was built on a 1.8-hectare estate at Nassim Hill. Neighboring Eden Hall were large homes built in the European style, such as Clovelly, Fairlie, Belvidere, and Neidpath, the latter the site of meteorological observations made in the late nineteenth century. One of the most famous homes of this period was the house of Hoo Ah Kay, benefactor of the Singapore Botanic Gardens land at Bukit Timah and himself the owner of a large estate and gardens to the east of Singapore off the Serangoon Road.[19]

When the settlements became too oppressive, the better-off inhabitants also had the option to relocate temporally. On Penang Island, especially after the arrival of the motorcar, the wealthy could circumnavigate the island, stopping off at beauty spots along the coastline or Tanjong Bungah in the north, home to the Penang Swimming Club. In Singapore, government-owned beachside bungalows were rented to civil servants for holidays. Some of the most popular were at Changi, where Europeans could "escape . . . to this shady verandah, these lofty cool rooms, looking over calm waters to green islands."[20] The beaches at Pasir Ris, Tanah Merah, and Pulau Ubin were popular for swimming and relaxing. Swimming clubs, such as that at Tanjong Rhu, were an option, as was owning or renting a bungalow or beach hut at popular Tanjong Katong for long weekends. Some contemporaries noted how, if these options were not possible, then "by resorting to the neighbourhood of the jungle a degree at least of reduction in the temperature may be secured" and that "sea bathing is also a relief within easy reach."[21] If one had to stay in town, then various methods were suggested to manage the hot days such as avoiding too much strenuous exercise, meat, strong drinks, and cold water. Eating cooling foods, such as salads, was especially recommended.[22]

Hill stations were also developed in Malaya, albeit somewhat later than in India, Philippines, and Indonesia.[23] A change in climate from the towns to the cooler hills of central and western Malaya was considered vital for the mental and physical health of European residents.[24] Although the Malayan stations were sometimes used as places of long-term convalescence, here they did not tend to double as medical research stations, as was often the case in India.[25] The first hill stations in Malaya were established on Penang Island (1830s) and at Taiping, Perak (1890s), but they consisted of little more than a few houses and roads.[26] New stations at Frasers' Hill and the Cameron Highlands, Perak, were developed in the 1920s. A "period of rest at one of those bungalows was a rest indeed of mind soul and body and to sit on the bench in front of Maxwell bungalow and look to the west over the Straits of Malacca at sunset was inspiring," wrote the Reverend Peach recalling his time at Taiping hill station during the twenties.[27] At the time of his recollection, the bungalows were restricted to European residents only, but around 1930 these restrictions were lifted.[28] It is worth mentioning here that Sir George Maxwell was involved in pushing ahead with the development of the Perak stations, the same Maxwell who was a key player in establishing momentum for the MMS during that decade.[29] No doubt his interest in climate played a role in both decisions.

It was during the interwar years that technology began to play a larger role in the cooling of Malayan inhabitants. This marked the beginning of a shift, albeit a slow and unevenly applied one, toward controlling—rather than simply managing—heat through new technologies. Although controlled atmospheres were normally small in scale or isolated in space—a cinema auditorium, a railway carriage, or a hospital—they relied on a wider infrastructure. The technology relied on local or national electrical infrastructures or capabilities to ensure a power supply, education to train local engineers to install and maintain the systems, and an import and sales network for local markets supported by a combination of private and public enterprise. Thus, controlling atmospheres, even within a domestic setting, required a centralized, top-down, and systematic mechanism that created an interdependent relationship of political-economic control and comfort.[30] Here then, the advent and application of air-conditioning on a mass scale would not have been possible without the foundational infrastructure of local electrical generators and, later, local and regional power grids.

POWERING MALAYA

The first consumers of electricity in Malaya were industrial, commercial, and governmental entities, essentially those with the funds and space to purchase, install, and run a generator. In George Town, for instance, some of the first electrical installations belonged to the government, with an electric telegraph in 1873 that connected Government House on Penang Hill with the town and a municipal street lighting plan vaunted (though not immediately enacted) in the 1880s.[31] Singapore, as the administrative and trading hub of the Straits Settlements, as well as the largest town, had the lion's share of electrical connectivity when it was first introduced. Following the US example, arc lighting was installed during the late 1870s at the Tanjong Pagar Dock in Singapore to facilitate night working.[32] Harsh arc lighting had little use outside of outdoor maritime or industrial settings, however, and thus the novel incandescent electrical bulbs patented by Thomas Edison in 1880 spawned far more interest. The position of assistant-protector for the Chinese in Singapore (a role created to enable better relations between the British government and the Chinese communities) was at that time held by a man named Nicholas Dennys. Outside of his day job, Dennys was also an inventor, and the creator of the Dennys-Cuff electrical lighting system, which he exhibited in Singapore in December 1883.[33] Linking his own house at Dalvey Road to the new system, Dennys invited around one hundred local notables and people connected to the telegraphic industry to showcase the system in use at his home as a marketing event. "The steadiness and brilliancy of the light evoked great admiration," a writer remarked in the local press, "and the simplicity with which the lamps were turned on and off seemed at first incredible."[34] The system comprised an automated battery-generated electrical current linked to Swan incandescent bulbs. A few days later, His Excellency the Governor and most of the municipal council members were invited to a subsequent exhibition, which purportedly was also highly successful, according to all accounts, and also according to the fact that the system was patented in early 1884.[35]

Dennys was not the only person promoting the wonders of electric lighting in Malaya. Another key figure in Singapore was Municipal Commissioner E. J. Wells, whose original role in Singapore was as engineer for the Gas Company, in partnership with Riley, Hargreaves and Company, which, perhaps somewhat ironically in the context of this chapter, was also an early investor in an ice-making company in Singapore.[36] It seems that Wells saw the potential in electrification, and he impressed upon the people of Singapore and even across Asia all the benefits of electricity over gas. He was responsible for some of the first commercial incandescent electric bulb lighting installations (the first was probably the Telegraph Office in 1885), including a commission by the Union Hotel in Hill Street.[37] Eleven incandescent electric lamps replaced the fifty-seven gas burners that had previously lit the music hall, the orchestra, the bar, billiard room, verandah, and entrance to the hotel in a wonderful advertisement of economy and taste, or so said the newspapers.[38]

In George Town, electrification took place about a decade after Singapore, but a dramatic display of electrical domesticity of a public "electric light reception" hosted by the Penang Club in 1901 helped persuade the town's elite of its benefits. Sponsored by the Municipal Council, "The Clubhouse was lighted with 16 and 8 candle power electric lamps," a reporter gushed, "and a powerful arc lamp was also exhibited . . . guests were taken . . . and shown the working of the different electric lights and the difference in the power of the lamps. Some kerosene lamps were lighted to show the difference between oil and electricity . . . they expressed their keen desire for no delay in getting Penang town supplied. All said they would give orders for electric lamps in their homes."[39] After the Municipal Council enabled a mains supply to be installed, many homes began to be electrified from 1901.

The electric Graham Bell telephone was brought to Malaya in 1879. This was the result of a private investment by Bennett Pell, who installed the first system within the Singapore-based Telegraph Company of which he was then manager. One year later, John Fraser and Company bought him out and ran the exchange on Princes Street until the Oriental Telephone and Electric Company bought them out, the US/Anglo-Indian venture dominating the field in Asia.[40] Singapore was connected to Johore on the mainland by the 1890s, but there was no international service until the 1930s. Looking at the phone company clientele during this period, clients were largely trading and commercial companies such as Boustead and Company or Guthries (import/export, shipping, mining, plantations), banking firms, or maritime organizations such as the Peninsula and Oriental Steam Navigation Company. The exception was the wealthy, powerful, and extravagant maharaja of Johore who installed his own personal line between Johore and Singapore the 1930s.[41]

Small-scale electrical installations could be run on generator power, but larger projects meant scaling up. The decision to introduce electric trams into Singapore, for instance, necessitated the city's first electrical power station to be built between 1902 and 1905. This powered the first regular tramway route, which linked some of the town's most iconic tourist and elite destinations: Raffles Place and the Esplanade.[42] The station also powered municipal street lighting. Electric motors began to replace older forms of power generation. The Cable Depot, center for the underwater telegraph cable between Singapore and other parts of Asia, installed electric motors in 1911.[43] But although industrial, commercial, and governmental parties were leading the vanguard, other types of organizations took up electricity too. In 1893, for example, theatergoers were treated to an amateur dramatic adaptation of the French play *Feu Toupinal*. This charity performance was memorable not for its treatment of the play but for the fact that it was the first stage performance in Singapore to be lit by electric lighting. Unfortunately, somebody forgot to switch on the footlights during the first half. The Armenian Church, first established in 1835, was the first church in Singapore to have electrical lighting and fans fixed in 1909, sponsored by a private donor.[44] In the early 1900s the number of electrical engineers advertising their services in the press blossomed, and the YMCA began running classes in electrical engineering, supported by government grants, in 1911.[45]

The electrification of Malaya was accompanied by an enterprising spirit. Thinking back to 1930, for example, a civil servant, T. P. Coe, recalled invitation from the owner to visit a small pineapple canning factory with the *Mentri Besar* (state minister) for Johore, Datoh Mustapha bin Jaafar. Down a tiny, deserted road about nine miles outside of Skudai (a small town in Johore), Coe and Datoh bin Jaafar found themselves in a well-kept village of around a dozen shophouses. "Each of these houses had electric light and electric fans" generated by a small hydroelectric power plant utilizing water from a hill stream, which also fed a reservoir. "A good example of private enterprise unassisted but unimpeded," Coe drily noted in reference to the colonial service, which rather adroitly also sums up British rule in Malaya at that time.[46] In 1930s Kuala Lumpur, the head of the Electrical Department told a slightly different but no less innovative story to the Reverend Preston Peach, an American missionary, about the introduction of domestic electrical appliances in Malaya. New electrical cookers, refrigerators, and other kitchen appliances had become popular across the country for the middle classes, he recounted, their rental regulated by the Electrical Department. However, the department soon found that the rent of cookers actually outpaced the sale of electrical power, which they found confusing. On checking all the homes where stoves had been rented, it turned out that many buyers had the cookers in their kitchen but had not connected them to the supply, preferring instead to continue using traditional methods of open-fire cooking. They were there for display only. After this revelation, the Electrical Department insisted that a minimum charge for electricity must be paid at the same time as the rental of the appliance, to encourage people to put the cookers to use.[47]

In 1936 a massive power outage in Singapore revealed the extent of this newfound dependency on electricity, as cookers, refrigerators, and lighting all shut down. Trolleybus services, cinemas, and clubs had to give out refunds, and about three-quarters of Singapore's population had to go about their evening "by the light of candles, oil lamps and electric torches." In the Swimming Club "women powdered their faces and combed their hair by the light of matches," and men found that they "had the wrong trousers or coats on." Governor Shenton Thomas had to "feel" his way around Government House "until peons arrived with electric torches," and "candles stuck in beer bottles provided a backwoods setting in a few European homes."[48] With the exception of hospitals, transport, and some public street lighting, most of these "newsworthy" stories seemed to be accounts of mere temporary inconvenience affecting a particular swathe of the population. Nevertheless, such stories reveal the sense of potential and the promise of a more connected and technologically advanced Malaya, and of course, the electrification of the towns was a necessary precursor to controlling atmospheres.

COOLING SOLUTIONS OR ASPIRING TECHNOLOGIES

The modern space was one that was cool and pure. When American air-conditioner manufacturers entered the Asian market, they brought with them the certainty of Western science, from studies undertaken during the 1920s and 1930s on the health benefits of conditioning and cooling air that may have appealed particularly to expatriates used to colder climes.[49] This was not at all the same as the benefits of the portable domestic electric fan, invented during 1880s, that offered the well-to-do home or modern office a new means of reliable, mechanical air flow.[50] "Technical hygiene" was the aspiration of interwar Malaya, at least according to press advertising and marketing, and at least for anyone who could afford it. Although the germ theory of disease in scientific medicine in the early 1900s had far overtaken older conceptions of the unhealthful tropics, ideas about the effects of heat on the (European) body persisted and intermingled with newer concerns about industrial pollutants, dust, and allergens in the rapidly growing colonial towns, especially after the introduction of the motorcar and its associated infrastructures. Air-conditioning became the pinnacle of achieving the sanitized home or office and thus became aligned with the principle of modernity.[51]

In Singapore, the introduction of air-conditioning was hailed as an invention that would provide not only comfort but better health with Western technology. "In Europe and America," it was noted in 1936, "where researches and inventions move ahead of time, as long ago as 30 years doctors and scientists began a serious study of what makes people comfortable. Particularly was this so in America, where scientists definitely established air circulation as one of the principal requirements for comfort and health." Singing the praises of the Trane Air-Conditioning System, the *Morning Tribune* staff writer continued: "Singapore may be a beauty spot and its strategic position superb, but its climatic conditions have always been a moot point with residents and medical men alike. Its humidity or dampness is the sore point, but the marvels of modern science and invention can transform the air condition of the interior of Singapore's buildings to the most delightful and healthiest obtaining anywhere. . . . After years of experimentation, air-conditioned offices, and homes, and even railway trains and closed motor cars, have come to stay."[52]

Benefits to workplace productivity were frequently used in pro-air-conditioning narratives. In 1937, for instance, "Definite Benefits to All" were touted by the *Malaya Tribune*, with air-conditioning making "it possible for men to work where the temperature previously rendered it impossible," a boon for keeping down the incidence of airborne diseases such as tuberculosis as well as pollutants.[53] Two years later, a study on air-conditioning in the tropics was reproduced, also in the *Malaya Tribune*. Written by R. E. Thompson, the article was largely based on American reports and statistics. While praising the "valuable contribution" in improving mortality outcomes in hospital surgeries and in nurseries, Thompson also spoke on the public health value of reducing the effects of outdoor air pollution by cleansing indoor air.[54] In 1938, the writer of another article sought to address what he termed "ignorance" surrounding the technology. Noting how air-conditioning tempered indoor odors, smoke, and dirt, health and cooling were not the technology's only function. "Putting the description of air conditioning briefly," he began, "it is the simultaneous control (in an enclosed area) of the temperature, the humidity, motion and cleanliness of air." Air-conditioning granted the ability to manage an atmosphere where there was "too much moisture in the air," meaning that "perspiration will not readily evaporate into air already heavily laden with moisture." The main purpose then of air-conditioning was to "remove the excess moisture," which was considered more of a problem than the heat.[55]

This argument would have had resonance in the tropics where the wet bulb temperature was held as the chief cause of heat-related mortality, rather than the temperature itself. This was the reason, Professor J. W. Gregory had written for a sectional address of the BAAS in 1924, that the humid regions of the empire were more dangerous than any other regions to people of all races.[56] Indeed, this may be why, as Priya Jain has also argued for British India, that the anti-air-conditioning narrative that was apparent in America was manifestly less noticeable in Malaya, where there was a powerful legacy of ideas concerning humid, unhealthy, miasmatic airs.[57] In nineteenth-century America, Salvatore Basile tells us, the advent of air-cooling technologies had met with great scepticism. Not only were people unsure as to the whole point of cooling air in the first place (heat was a fact of life), but early experiments did not work well and were thus often considered little more than a gimmick.[58] In Malaya, however, there was little public discussion concerning air-conditioning until it had become a more viable technology in the twentieth century, and although there were the occasional critics, their arguments were relatively mild—perhaps suggesting that moving so suddenly from warm to cold, or vice versa, might harm the body, thus advising caution during the transition between hot and cooled airs.[59]

Mechanical air ventilation systems for buildings were the first iteration of this transition to modern cooled and healthy spaces.[60] In Singapore, Kartar Singh, for example, recalled how his father helped build the new Capitol Theatre when works started in 1929. Developed by the family-owned firm M. A. Namazie and Sons, the theater and cinema space design was influenced by that of New York's Roxy Theatre. Incorporating a heat-reducing scheme whereby air was circulated around the theater from a suction fan in the basement, the air then passed through a ventilation system that filtered, purified, and "washed" it before "electrically" charging it in an Ozonair machine and pushing it out into the auditoriums through air vents, "with all the freshness of a cool breeze."[61] The Ozonair pure air ventilation system was patented in London in the early 1900s, but the company also manufactured components used in conjunction with other systems.[62] Within a decade, this system was beginning to look old fashioned.

The Alhambra cinema is considered to have been the first cinema in Malaya to have a *modern* system of air-conditioning, installed in 1938.[63] The Alhambra was at that time owned by a local entrepreneur, Tan Cheng Kee, who owned two other well-known theaters: the Pavilion on Orchard Road and the Marlborough on Beach Road. In 1938, Tan leased the Alhambra to the Shaw Brothers Company (comprising Runme and Run Run Shaw), which enterprise now dominated the Malayan entertainments industry, operating around sixty cinemas and four amusement parks across the peninsula. Tan upgraded his lavish theater spaces in the early 1900s with electric lighting and fans, but the new coalition with the Shaw Brothers meant the now elderly Tan was able to see the Alhambra totally refurbished. The air-conditioning installation necessitated extensive restructuring of the roof as well as the introduction of new air ducts.[64] The new system attracted the most attention of all the renovations, despite the new bar, seating, and décor and dominated the discussion of the opening ceremony attended by Governor of the Straits Settlements Sir Shenton Thomas. "Entertainment in the East is controlled for the most part by the problem of weather," a journalist noted in the *Morning Tribune* on the day before the grand opening. "It is a remarkable fact that the idea of installing an air-conditioning plant has not been thought of before."[65] In reports of the event the following day: "There is no doubt that the air-conditioning is a huge success" stated the writer who spoke on the new system before any other aspect of the refurbishment, "all through the evening there was a pleasant sense of cool dehumidified air . . . and there was no suggestion of any intolerable heat that can so often ruin an evening's entertainment."[66] The new system even attracted notice from overseas, with the manager of the Maxim Theatre in Surabaya—Java's only air-conditioned theater—sending a telegram of good wishes to the New Alhambra "from one air-conditioned theater to another."[67]

The late 1930s saw a proliferation of public spaces incorporating air-conditioning into their design. New World Park, for example, was an entertainments space that occupied an area between the Jalan Besar and Serangoon Roads between 1923 and the late 1960s. The park comprised amusements, a cinema, cabaret, shops and restaurants, an opera and a dance hall. Dance halls had become popular after the First World War in many big cities across the world. Some were elite spaces, such as those described by Andrew Field in 1920s Shanghai, but others were less exclusive, offering the working classes a new social outlet.[68] Like the elite spaces of Shanghai (in fact, likely modeled on these given its ownership by the Shanghainese Shaw Brothers, while catering to a slightly wider audience), from 1938 the New World Dance Palace also offered the exclusive opportunity to "Dance in Perpetual Coolness."[69] As James Nott has argued, the dance hall emerged as a distinct social and architectural space that embraced modernity and affordable luxury of a type previously reserved for the elite. A significant part of this modernity was design, in line with new consumerism and technological innovation.[70] In the tropics, air-conditioning was a critical part of this claim to modernity. According to the *Morning Tribune* in May 1938, the New World Dance Palace and its cabaret hall was "destined to be the most modern and luxurious in the whole of Malaya," with a "new air-conditioned building [where] 1,500 people . . . can dance in a cool, dehumidified building." Here, all the glaring electric bulbs were carefully hidden behind opaque glass; there was a revolving stage and above the dance floor was suspended a "magnificent chandelier of striking, futuristic design"; butdespite all this, "it will be the air conditioning that will intrigue most people."[71]

The New World Dance Palace has been described as a site of experimentation and "crucible of colonial modernity . . . at the nexus of modern consumption and mass culture."[72] One might take this claim a step further and argue that the technological capacity to control the atmosphere lay at the heart of this consumer-orientated venture. Headlining many of the articles of the day and used as a major marketing ploy, the ability to dance within a climate-controlled environment was a major innovation. As an advert for the new domestic Frigidaire air-conditioning unit declared in 1937: "In the Tropics . . . Air-conditioning is not just a passing fad—it is a vital contribution to the health and comfort of the modern home . . . on a par with electrical refrigeration and modern sanitation."[73] Kenneth Black, formerly senior surgeon at the Singapore General Hospital and professor at the Medical College, stated a similar opinion in 1935, when he argued air-conditioning meant "that the dweller in the tropics would never be exposed to the oppressive heat except in the open air."[74] And this was not only about cooling but also about controlling pollution such as cigarette smoke. Marketing campaigns for entertainment spaces like the New Alhambra or the New World dance hall focused on removing smoke irritants so that customers could watch or dance in fresher air as well as cooler air. This was a key claim of air-conditioning companies. If you used the latest Carrier air-conditioning technologies imported from the United States, at least 98 percent of air impurities would be eliminated—or so the adverts claimed.[75] The designers of Singapore's Supreme Court Building in the late 1930s may have had this in mind when they chose the SoloAir System of ventilation, supposedly patented in Singapore by E. H. Hindmarsh, to supply the new classical-style building. A collaboration between the Public Works Department and United Engineers, the new building was constructed over three years, completing in 1939. The SoloAir System of ventilation was designed to provide a means of removing dust and pollution from the air as well as cooling it. The system was relatively simple, with air drawn through filters and the propelled through a series of ducts using centrifugal multivane blowers, manufactured by Ozonair Company Ltd of London, who had supplied the original mechanical ventilation system at the Capitol Theatre.[76]

Ten years after Kartar Singh's father helped build the Capitol, the premises had a major fifty-thousand-dollar refurbishment during which a new, modern air-conditioning system was installed as well as a concrete projector room, a building material that had been newly introduced to Singapore and was the epitome of modernist construction.[77] Adding air-conditioning to the theater building had not been straightforward, however. It involved redesigning much of the interior, removing fans and ventilation shafts, and adding in a "cooling tower" to remove heat in the roof. Like many systems of the era, it was also designed to extract the vast amount of cigarette smoke that was generated in all public buildings at that time replacing the smoke with around one ton of cooled air every minute.[78] The job was undertaken by the firm United Engineers, who were Singapore-based contractors for all types of government and commercial air-conditioning installations of that era, among other electrical work. United Engineers were the leading British engineering firm in Malaya, which had devolved out of two older companies, first Howarth, Erskine and Company Ltd, a company that quickly broached the electrical market and was hiring electrical engineers as early as the 1890s. This company then merged with Riley, Hargreaves and Company, eventually becoming the United Engineers Ltd in 1912.[79] They worked with equipment supplied by the Carrier Air Conditioning Company of America, which had entered the Asian market in the early 1930s eventually setting up a regional headquarters in Singapore, alongside Tokyo, Hong Kong, and Sydney.[80]

At first, United Engineers were the sole representatives for Carrier, but as trade picked up, from 1936 the firm moved its own marketing representative to the Malayan peninsula.[81] The first Carrier dealership was established in Kuala Lumpur by 1948, and four years later another dealership was set up in Penang, selling twenty-five room air-conditioners within its first year of operations.[82] United Engineers and Carrier were instrumental in making Malaya an air-conditioned nation. Combined they were responsible for cooling the air in the first medical installation in Singapore, the operating theater of Singapore General Hospital in mid-1938, but interestingly also Singapore Dairy Farm, where air-conditioning for their cattle sheds was introduced the same year. During the 1920s the Cold Storage Company, a frozen food import chain that had been established in 1903, diversified into producing milk and dairy products, including ice cream.[83] This line was rebranded in 1937 as "Magnolia," a company that still produces dairy products for modern-day Malaysia and Singapore. Magnolia's base was the area known today as Dairy Farm, within the modern Bukit Timah Nature Reserve. Run by British expatriate Fred Heron, one of its goals was to provide children in Singapore with pasteurized milk, emulating the social welfare drive then at its apex in Britain, the "Milk in Schools Scheme."[84] One of the reasons that Dairy Farm's imported Friesian cows did so well outside of their native temperate environment was the addition of air-conditioning in their stables, helping to keep their temperatures at the optimum level for sustained health and fertility.[85]

Many of these air-conditioned and cooled spaces were repurposed during the Japanese occupation of Malaya during the Second World War. The Capitol Theatre, for example, with its massive, recently installed Carrier system was used as a warehouse to store flour. All the seats were taken out to accommodate huge quantities of dried goods, and the upper rooms, previously a bar, restaurant, and dance space, became a mess hall.[86] The air-conditioned space of the Alhambra Theatre did not suffer the same fate but continued as a site for screening films throughout the war, specifically films created by Japan's propaganda industry.[87] In 1941 Tan Cheng Hwee had started working at the frozen food import company, Cold Storage. He was there for only three weeks before the British surrender, and the store continued to stay open throughout this time, despite a bomb falling on the kitchen department. After a brief hiatus, the Japanese began hiring staff again, and as a former employee Tan easily got his old job back. Due to the refrigeration much of the frozen meat and vegetables that had been brought in from Australia prior to the surrender was still usable, although some of the storage facilities had been bombed. According to Tan, this amounted to at least twenty-six cold rooms. The food was reappropriated for the Japanese military and for hospitals.[88] Thus air-conditioned space became a vital part of the war effort, enabling an element of food security at a time when imports were critically damaged. With the repurposing of cooled space for the Japanese military, patterns of access to controlled atmospheres followed the British colonial trajectory, starkly revealing that atmospheric control was a function of unequal power relations in (any) colonial society.

ASPIRATION VERSUS ACCESS

In 1938 Singapore's *Sunday Tribune* reported on a speech given by an electric company official in the United States. Selling the potential of electricity—from futurist domestic appliances to germ-destroying ultraviolet—the company representative went on to state how, by the 1960s, man would "live in a house lighted, cooled, humidified and air-conditioned by electricity."[89] The Malayan Exhibition, organized by the Malayan Agri-Horticultural Association and held in Kuala Lumpur one year later, likewise placed cooling technologies at the heart of a future-thinking agenda, featuring an air-conditioned room as part of its displays and exhibits, which (unsurprisingly) was popular with visitors who enjoyed the "cool compared with the humidity and heat outside." Many asked about the cost of installing the equipment in their homes or offices.[90] Superficially this room may have been an anomalous exhibit in a show aimed at showcasing local agricultural and horticultural expertise, but the inclusion of air-conditioning represented the future of technological intervention in farming as much as in any other realm, as the Dairy Farm example reveals.

By the 1930s the sense that science and technology were not simply part of the future but part of the now had infiltrated popular culture.[91] This reverse teleology—the conceptualization of a steady march into a technological future where cities were fed by filtered air and artificial sunlight—penetrated into the heart of the tropics, even if such "Wellsian" futurism was scoffed at as being fantastical.[92] The old classical style of architecture was "giving way to a type more suitable to the tropics," declared a writer in 1938, with "modern, flat roofed, reinforced concrete" lending more coolness than tiled roofed designs, the new concrete box promising the ability to do away with the mosquito net and natural air flow now that the air-conditioner had arrived.[93] Such hypermodern homes—such as those then being built for the wealthy Chee Guan Chiang on Grange Road or for L. Langdon Williams at the Singapore Golf Club—epitomized the cool future of housing in Singapore.

In Malaya, as elsewhere, domestic air-conditioning has been beyond the reach of many low-income households until relatively recently, not least because of the price of the unit, its installation, and running costs but also because of basic inconsistencies in the electrification infrastructure for homes in deprived rural and urban settings.[94] Priya Jain has argued that in colonial and postcolonial India, air-conditioning remained the privilege of the wealthy for many decades after it was introduced. Its uptake was complicated by the fact that the secondary benefits that air-conditioning afforded (less domestic dust, for instance) were not so important in a place where cheap domestic labor was easy to come by.[95] This was most certainly the case in Malaya as well, where domestic labor was ubiquitous and little better than indentured slavery in many cases.[96]

Here too, there was an uneven adoption of electrification in villages and towns. Ronald Milne, for example, spent his childhood in a Eurasian community in Upper Serangoon in Singapore. His father was a master mariner who worked for Teo Ho Lye and Company, which ran a boat between Singapore and Sumatra dealing in tobacco. They were not connected to a water supply and lived off well water for his entire childhood, but they had electric lighting and a refrigerator by 1929. This meant that, for the first time, he was able to experience iced water.[97] Niranjan Singh, who was born into a Sikh community in India, came to Singapore at the age of four. He described his family as "middle-class." They lived relatively near the Milnes at Hillside Drive, which at that time was a newly developing suburb on the outskirts of town, and he had electricity from 1933.[98] But Chan Kwee Sung, whose father was an unregistered dentist in Singapore, used oil lamps, as his family did not have electricity in their shophouse on Tanjong Pagar Road where they lived during the 1930s. Mehervan Singh, also a longtime inhabitant of Tanjong Pagar, recalled that his house did not have an electric supply until 1960.[99]

When it came to air-conditioning, a press article from 1937 suggested that, although the technology had been around for some ten years or more, most people were shy of getting it installed in their homes or offices, which the correspondent blamed on the "conservatism" of Malayan residents. However, the article also noted the cost of around six hundred dollars per domestic installation, which may have had more influence on the matter. This technological inequality is aptly described by events in 1926, when a heat wave saw "hundreds of Asiatic residents from the town districts . . . spending their nights in the open, the Esplanade having been the centre of unusual crowds during the evenings and nights for the past few days," a stark contrast to the wealthy in society for whom it had been "impossible to get a cold bath without the addition of ice, the water when turned on from the taps being quite warm."[100]

Likewise, in 1948, another heat wave sent people scurrying to the beaches at Katong and Changi, and "while town dwellers who were fortunate enough to have a roof garden spent the night on camp beds in the open, others spent the night on five-foot ways, verandahs and balconies."[101] Kartar Singh, too, while talking excitedly about air-conditioning in his recollections of the prewar Capitol Theatre, observed how most of the patrons were British or European expatriates, though there were cheaper seats available for around fifty cents that local children occasionally paid for. Most of the films were American or British and only once was an incredibly long South Indian film shown, starting at midnight. As to his own moviegoing experience, Singh could not afford the films, but he knew about all the goings on at the theater as he and his father lived there for a time after his mother's death. He therefore enjoyed some of the air-conditioned hours within the theater space by default, although he noted how all their meals were cooked on the street outside.[102]

Other everyday forms of cooling technology became more widely available during the 1930s. In the larger urban areas, the choice and quality of food, for example, improved with the rise of refrigeration and refrigerated transport. The Cold Storage Company (whose flagship Orchard Road store was of modernist design) imported frozen meat and dairy from Australia and acted as a storage and distribution service for fresh goods across Malaya, especially after the extension of the refrigerated rail network across the peninsula.[103] By 1914, the service was extended out to Penang, Ipoh, Kuala Lumpur, Taiping, and Klang; by the 1930s, refrigerated train services meant that a variety of fresh goods was transported across the peninsula from Europe, as well as Australia.[104] The advent of the supermarket chain transformed the diets of many wealthy inhabitants, who were often disinclined to eat much of the locally produced foods, preferring fancy imported goods or, simply, brands that they felt more familiar with. As Bruce Lockhart, a rubber planter, was to declare in 1935: "Singapore gets fresh meat from Australia, fresh butter from New Zealand, swede and turnips from Sumatra, potatoes from Palestine, tomatoes from Java, rhubarb from New South Wales, oranges from China, and a dozen other items such as Edam cheese and Droste chocolates from Holland."[105] Paul Carroll noted how in the 1930s his mother would buy seafood such as prawns and shrimps from the Cold Storage as they were considered better than the goods sold on the open market where the "locals" would shop, revealing how refrigeration was marking a significant divide between the expatriate and wealthy communities and other inhabitants.[106] The frozen food chain spread to all the main towns on the peninsular, including Penang and Kuala Lumpur during the 1930s.[107] George Caldwell recalled how Cold Storage supplied the big hotels but also Boon Bee, a Chinese owned importer that specialized in fancy goods for wholesale, such as smoked salmon and French cheese, and an Australian company known as Fitzpatricks, which was eventually bought out by the Cold Storage chain.[108] Cold Storage, Boon Bee, and Fitzpatricks were not for everyone. Ordinary people did not shop there, as Teo Yew Chiat recalled; they only worked there, just like Cheng Hwee Tan at the outbreak of war in 1941.[109] As a visitor to the Sea View Hotel remarked in the 1930s, expressing this clear class divide: "here was a revolution . . . cold storage, electricity, the automobile, and air conditioning for Singapore's plutocracy—all inventions that removed life's main discomforts in the tropics."[110]

Even for commercial ventures and for government or publicly funded institutions, air-conditioning was unevenly appropriated. Not all cinemas were air-conditioned until the 1960s, for example, and many schools lacked any form of cooling technology, even fans in some cases.[111] Brother Joseph Kiely, for example, moved to Singapore in the 1950s to work and live at the newly built Brother's Quarters of St. Joseph's Institution school on Queen Street in central Singapore. He recalled how only the Community Room had air-conditioning and there were no fans in their individual quarters, hence he spent a great deal of time in the Community Room![112] Recollecting his time in Singapore from 1962 to 1965, Royal Air Force officer David Penberthy told how air-conditioning barely existed in Singapore. Until you acclimatized to the humidity, the only way to get to sleep was to have "two or three pints of Tiger to help you do so." He could not recall air-conditioning in "any of the private homes that I knew of, with perhaps one or two exceptions in the newer luxury apartment blocks that were going up at that time . . . in other words, we lived in exactly the same way as the local population lived."[113] His recollections suggest that the borders were murky between who had and who had not adopted cooling technologies.

Indeed, according to Carrier's records in the Far East, in Kuala Lumpur, Carrier units were installed in only a few commercial and governmental buildings, including the Federal Teaching Hospital, the Straits Times Press (Malaysia) head office, and the Chartered Bank, by the 1960s.[114] The situation was similar in Singapore, with governmental ministries and embassies being the beneficiaries of large-scale Carrier installations with the more diversified list of installations—such as in restaurants, hotel chains, shopping malls, or in public housing projects—not taking place until the 1970s.[115] This was likely the case as even public libraries were only partially air-conditioned despite many new branches being newly opened in the 1970s. As Chan Thye Seng, a public servant who was then involved in expanding Singapore's library service, recalled, their high-ceiling design was meant to assist in air flow within the sections that were not artificially cooled.[116] In many cases, air-conditioning was not adopted until well into the 1990s and beyond, a fact that related more to the Malaysian and Singaporean economic situation than to any defect of the technical imagination.

COOLING AIRS, SOCIETY, AND THE TECHNICAL IMAGINATION

David Arnold has suggested that one of the characteristic features of Western technology was its claim to universality but, if left unmodified on transposition into new contexts, this universality had a high potential for failure.[117] His claim could be juxtaposed with the Malayan context, as the uptake of cooling technology was situated more in nature, context, and circumstance. Sectors where air-conditioning or refrigeration had evident benefit, such as at hospitals or as part of commercial food supply chains, were obvious success stories, but for everywhere else, the benefits and the costs were not so clear. In 1979, Frank Trippett, a columnist for *TIME Magazine*, queried why scholars had ignored the sociocultural significance of the introduction of air-conditioning when the implications of other technological advancements—such as the automobile and television—had been well covered. This snippet, sourced by Raymond Arsenault, began an important early article on the sociocultural impacts of air-conditioning in the American South, arguing for the central role of the technology in changing aspects of life in hot southern states, for better or for worse.[118] Singapore's most famous prime minister, Lee Kuan Yew, would likely have agreed with Trippett, famously giving air-conditioning credit for having "changed the lives of people in tropical regions" enabling "better concentration and productivity" in a speech delivered in 1999.[119]

There has been no smooth transition into a cooled indoor future in Malaya though. Technological infrastructures do not operate as inert things but are instead products of socioeconomic and political circumstances.[120] To think of them in this way enables us to consider not simply the circumstances of their production but the people and realms they touched. The tropical environment gave every reason for cooled indoor spaces to be hugely popular, but as Arnold argues, the natural environment was less important than the political environment within which the technology operated. This environment was deeply unequal. In an era before the term "sustainability" became a buzzword, electricity and air-conditioning offered a new technological vision of the future. But this imagined future was not for everyone, and in the late 1970s, air-conditioning remained a fictional means of describing the hypermodern indoor environment in a new wave of Singaporean science-fiction authorship.[121] It was only when the economy of Malaysia and Singapore strengthened in the 1990s, and more people began living in condominiums or apartment-style government housing (known as "HDBs" in Singapore), which design compelled indoor cooling, that the situation shifted for many. It is ironic that just as air-conditioning has become the norm for most people, urban heating has escalated, with the conditioning units churning out more heated air onto the streets and back lanes of today's Singapore and George Town, resulting in a thoroughly modern catch-22.[122]

CONCLUSION

THE DUSK OF EMPIRE

In 1963 a report was submitted to the Conference of Commonwealth Meteorologists, held at the United Kingdom Meteorological Office Headquarters in Bracknell, about the immediate two-year future of the Malayan Meteorological Service (MMS).[1] Following the first step of independence from Britain, the Malayan mainland and Singapore Island functioned under the umbrella of the Federation of Malaya, with Singapore operating as a self-governing state from 1959. In 1963 Singapore, along with Sabah and Sarawak, merged with Malaysia to become the Federation of Malaysia until 1965 when Singapore forged an independent state. These years of uncertainty and transition are reflected in the history of the MMS. From 1963 to 1965, although Singapore was run as a self-governing state, responsibility for research was placed under the Federal Constitution of Malaysia, not at the state level. This makes some sense as defense, which had been a significant part of the MMS role for the past two decades, was also a federal responsibility.[2] The service operated between main sites in Selangor and Singapore and underwent a change of location on the mainland before finally splintering into the Malaysian Meteorological Department and the Meteorological Services Singapore in 1965. The former was placed under the Malaysian Ministry of Transport, run from Kuala Lumpur, and the latter under the Singaporean Deputy Prime Minister's Office, run from Singapore.

The 1963 report was issued during this period of flux, and its two-year predictive span reveals much about the state of service and science at that difficult political time. First, the commitment to improving tropical weather research that had been lacking in the earlier part of the century. Over the duration of British rule, Malayan meteorology was developed from an ad hoc affair to a fully operational and modern service with international representation, but the lack of funding always meant that priority went to forecasting. The report stated how the services were understaffed (for at least two decades), and it was hoped that the coming political changes would enable the recruitment of more scientific officers to undertake critical research into equatorial meteorology. The second point concerns the staffing itself. The report was written by then director of the service in Singapore, K. Rajendram, the first man to hold this post who was not British. He had started in the service in the 1950s, then worked his way up the ranks to become director on the retirement of Dr. C. A. Lea, and would lead the service in Singapore when it was reestablished in 1965 as an entity under the Deputy Prime Minister's Office.[3] Looking after a team of seventy-eight members of staff, Rajendram was responsible for shaping the new country's direction in equatorial meteorology, and he witnessed firsthand the transition of the service from its days on a postwar shoestring budget to a technologically advanced department with modern facilities in the early 1980s.[4] He also helped the service have senior representation on the global stage, serving as a regional association president of the World Meteorological Association (WMO) across the late 1960s and early 1970s.[5] His counterpart in Malaysia was Ho Tong Yuen, a previous director of the MMS in Kuala Lumpur and now director of the MMD after the two coutries split. He also represented Malaysia at the WMO.

The appointment of Rajendram and Ho are important as it reveals a major transition from the era of imperial meteorology: the employment of non-Europeans in key scientific roles. Although the service was manned by a predominantly Asian staff, many of whom were doing the observational work out at stations around the country, few if any made it to more senior positions. Kanagasabai s/o Narayasamy, who had been at the Singapore MMS headquarters on the advent of the Japanese occupation, for example, came back in 1948 to work for the service for three years. He recalled how roles like his—a weather observer making records—were unpopular as the staff "spent three-quarters of [their contract] in a station and the last quarter in Singapore."[6] This was typical of the roles held by non-Europeans, especially before the war. In the postwar era the pattern began to shift, however, toward employing Asians in senior scientific roles, and, indeed, all roles, in order to make the service entirely sufficient on local labor after the political transition.

In 1949, for example, K. C. Koruthu made the headlines as the first Asian officer to be employed at the MMS. Koruthu was university educated and had been with the colonial service since 1928, moving to the meteorological branch in 1935, mostly working at the Forecasting Office at Kallang.[7] He was at the Kota Baru station when the Japanese invaded from Thailand in 1941. From 1949 onward, he held the title of assistant meteorological officer.[8] Koruthu and Rajendram would have worked together in Singapore in the 1950s, along with Hwang Tiaw Sooi, who had trained in meteorology at the Australian Weather Bureau in the 1950s, something that would not have been possible for Malayans before the war, most of whom only went to local technical colleges.[9] Hwang progressed rapidly within the service, becoming Senior Meteorological Officer under Rajendram's leadership and flying out all over the world to represent Malaya at meetings of the WMO and the Commonwealth Meteorologists.[10]

Hwang, like Rajendram, was also part of the meteorological services' transition into modernity. In 1963, during a water shortage, he was the senior scientific advisor at the airport overseeing some of the first rainmaking operations over Singapore and Johore in Southern Malaysia under the guidance of E. J. Smith from the Commonwealth Scientific Industrial Organisation.[11] This cloud seeding operation, utilizing the RAF Butterworth air base and a DC-3 plane from the Royal Australian Air Force (RAAF), was operationalized after Singapore's inhabitants had been put on twelve-hour water rationing.[12] Taking inspiration from successful cloud seeding experiments across the world, especially in Australia, the experiment entailed a collaboration between the Malaysia Air Charter (MAC)—Malaya's main internal aircraft carrier), the Public Utilities Board (PUB, responsible for water supply in Singapore), and the Australian High Commission, with MAC taking over after the plane borrowed from the RAAF had to be returned.

Expert officials at organizations such as the PUB felt that a nuanced approach was necessary, one that took into account all possible solutions. The organization was established in 1963 to oversee all public utilities for Singapore, a state, not a federal, responsibility. It was seen as the answer to the silo mentality created because of the existence of separate authorities' jurisdiction over water—namely, the Municipal Council and the Rural Board. This hindered the realization of joined-up infrastructural solutions for water management. With a major drought overshadowing the first few months of operations for the newly inaugurated PUB and most of the country on strict rationing, it is perhaps no surprise that the chairman, Dr. Fong Kim Heng, felt the need to work with rainmaking expert E. J. Smith and Singapore's meteorological team to investigate every possible solution. The previous government's attempts to provide enough water to the populace and to manage through unusual dry spells had always been a game of playing catch-up and the new administration had inherited a system that was neither self-sufficient nor able to cope in the event of a significant dry spell.

The rainmaking experiment demonstrated a significant sociotechnical and scientific shift in thinking about managing extreme weather in Malaya that would have been unthinkable before the war, despite the interest in rainmaking experimentation around the world. Was this the future of drought control? Commentators were divided. "Science was ready to make it rain," stated the *Straits Times* on 9 May 1963, as that year's drought rolled on interminably.[13] But the rainmaking operation over which Hwang and Smith presided was continued in fits and starts over several months, with multiple failed attempts to find suitable clouds to seed with silver iodine crystals.[14] The protracted nature of the experiment attracted public comment, and not all good. Having an aircraft on hand all year round for "such an occasional job" was just not practical, argued Mr. A. J. Morris of Singapore. A better idea might be to have a loaded rocket on hand, designed to distribute the silver iodide at the exact moment a suitable cloud was spotted.[15] Not unsurprisingly, there is no record that the MMS took this suggestion seriously.

Just a few years later, in 1968, modern science and technology were called upon once again, this time not to ameliorate drought but to prevent flooding. An important seminar for meteorologists was hosted in Malaysia, just outside of Kuala Lumpur in Petaling Jaya, at the headquarters of what was now the new Malaysian Meteorological Department. Following a deeply divided and even violent period of unrest between the newly independent countries of Singapore and Malaysia, this seminar saw the department under Ho Tong Yuen working alongside Rajendram representing Singapore in the name of scientific cooperation.[16] The 1968 seminar, "Forecasting of Heavy Rains and Floods" (rather ironic given the coincidence of the meeting with a northeast monsoon–inspired flooding episode) was for representatives of fourteen regional services, all of whom were invested in the perennial problem of monsoon-related storms and subsequent flooding. Held in late November in Kuala Lumpur, the two-week event took advantage of the weather to see flood-warning systems in situ on a field trip and raised public and scientific awareness of the combined functionality of meteorology and hydrology.[17]

Describing the meteorologists who attended this meeting as "space age 'bomohs' [a Malay shaman]," Hugh Mabbett, a maverick Asia-based journalist who worked for many years in Singapore, speculated whether weather would eventually be tailored to "meet man's needs."[18] It was certainly the case that technological developments in the 1960s had led to an optimistic outlook on the near future of meteorology, Mabbett's statements were more in tune with the world of science fiction than science fact. Graham Sutton, director of the UK Meteorological Office at this time, argued how automation and computing were, and would soon, take over many of the slow manual aspects of meteorological research and forecasting. "Forecast charts will come directly from the computer," so that "the laborious plotting of charts by hand" will become a thing of the past, he anticipated in 1964. On the future of weather modification, however, Sutton cautioned that "there is no certainty that any method of modifying weather or climate on a large scale is likely to succeed, and no attempt now would be worth the risk and expense . . . weather modification must remain the greatest uncertainty of all."[19] There were no limits to science fiction, but still many to the realization of science fact.

Indoor weather control was far easier. Though the focus in this book is on indoor cooling technologies in largely domestic or commercial public settings, as Yuriko Furuhata points out indoor weather control—the creation of artificially cooled and dehumidified space—was also integral to the rise of modern computing and, thus, meteorology. The capacity to create a climate-controlled environment had facilitated the large-scale data crunching necessary for modern weather forecasting and numerical weather prediction.[20] The creation of a micro-climate for predicting the macro-climate presents an interesting trajectory of thought that benefited tropical Malaysia's and Singapore's meteorological services postindependence. Describing the MMD offices in Kuala Lumpur in 1968, Mabbett told how "a screen flickers with green light, instruments buzz and click . . . and a white coated technician operates a control board. Several times a day he steers, by remote control, an antenna on the roof directly at weather satellites passing overhead. The antenna picks up signals for the satellites and the other equipment turns them into photographs that emerge within seconds, still wet with processing chemicals."[21] The scene that Mabbett depicted is more akin to the bridge of the USS *Enterprise* in Star Trek (which had premiered in the United States two years earlier) than the nineteenth-century Palladian-style meteorological headquarters in Singapore that the British had founded.

Meanwhile, the new Republic of Singapore was likewise investing in new technologies. These were directed toward the more hands-on parts of the service operating from the airport at Kallang and, from 1971, at the Paya Lebar airfield. Opening a two-week seminar on synoptic analysis and forecasting in the tropics in Singapore in 1970, the communications minister Yong Nyuk Lin announced the imminent arrival of a new remote-controlled ground station to receive satellite data.[22] Satellite and infrared surveillance had dramatically improved the capabilities of the service to track storms and to forecast landfall or intersection with flight routes, in a way that was not possible even with the advent of radar sensing.

The introduction of weather satellites into world meteorology from the late 1950s worked in tandem with high-speed computing for analysis, numerical simulation, and modeling of climates. Much of this work was coordinated under the umbrella of the international effort known as World Weather Watch which had at its heart a trifold mission to improve global observation coverage and reporting, data-processing, and forecasting.[23] This was followed in 1967 by the Global Atmospheric Research Programme

(GARP), which undertook a series of collaborative research initiatives, including its monsoon subprogram and its field phases, one of which was the Winter Monsoon Experiment (MONEX).

In 1978 the Winter MONEX came to Malaysia. This research project was envisioned to monitor and gather data on the mechanisms that trigger the winter monsoon, with the aim of gathering enough data to improve the prediction of droughts and floods in monsoon-affected regions. In Southeast Asia, December sees high equatorial rainfall largely because of evaporation from the South China Sea carried by strong northeast trade winds, but the data available during the 1970s to understand the interaction between the planetary scale and more regional scales of the phenomenon was limited.[24] The winter MONEX was designed to address this. Hosting this experiment was a major undertaking, requiring an international operations center headed by Ho Tong Yuen at the MMD, in partnership with the WMO, using data received from satellites, weather radar systems, and weather balloons owned by Japan, Russia, and the United States, and research aircraft equipped with radiosonde probes and sensors, which flew over the China Seas and parts of Southeast Asia.[25] The first week of MONEX activities were hampered by poor weather, which meant that flying conditions for upper air research were unfavorable but otherwise, the experiments were considered a success.[26] More than eighty scientists from across Southeast Asia, including Singapore, but also from across the globe including China and Saudi Arabia, were involved directly at the Kuala Lumpur site, while others were employed at remote site observational programmes.[27] The new global weather satellite system was also expected to make possible both the monitoring of weather-related phenomena such as the distribution of ice and the observation of flood and drought conditions.

This technology was a far cry from the responsive engineered solutions of the earlier era, shifting the sole responsibility from the engineer into a shared realm with the scientist and enabling prediction and preempting, rather than simply reaction. This level of tropical climate research was also at a scale not yet undertaken, the latest technologies enabling gaps to be filled in the geographical dispersal of observations that colonial officers of the prewar era could only have dreamt about. Forecasting was still a major priority, but the furthering of understanding of global teleconnections and regional atmospheric circulation and phenomena was now coming directly out of Malaysia and Singapore, practiced by Asian officers in collaboration with international research teams.

The 1960s and 1970s were a period of deep infrastructural problems in both Singapore and Malaysia in terms of managing the impacts of the monsoon and flash flooding, so it is no surprise that their respective governments were intensely invested in monsoon research and improving predictions.

Nonetheless, while such projects saw global cooperation at a scale and technological capacity that could only have been imagined one hundred years earlier, flood control and drought prevention were still a matter of improving engineered solutions on the ground. A once-in-a-decade flood devastated Singapore in December 1978, for instance; it caused around seven deaths and was reportedly one of the worst to have affected the nation. Ironically, the floods coincided with the Winter MONEX which began on the first day of that month.[28] Solving the problem of water, in Singapore at least, has been described as a high modernist project by Loh Kah Seng and Michael D. Pante, who have described flood control as a political process, an expression of the dominant politics and technologies of the time, and a means of control of nature and thus nation.[29] Singapore's small size and rapid economic growth enabled comprehensive flood mitigation to be successfully incorporated into urban planning by the 1980s. But, more than this, a push back against disordered development, which included haphazard and reactive planning and policy that have formed the backbone of Malayan politics (and which have in this book been considered symptomatic of British rule). Where the British failed to control their environment, the modern nation of Singapore has succeeded.

Floods, tropical storms, heat, and drought have determined life flows for people living in the tropics from time immemorial, but colonialism drew sharper lines around who managed, and who did not, as people were increasingly moved out of traditional community coping structures. As indigenous forms of living with nature declined, Western or Western-adapted environmental and hydraulic management systems and domestic living styles took over. In many respects it is not possible to divorce Westernization from development and dominant capitalist structures, capitalism or economic growth having become a state to which most modern nations and individuals aspire regardless of a colonial past. The global narrative of technological intervention to control the environment is still alive and well today and is often considered the solution for arresting climate change—instead of simply cutting back on emissions, for to do that means tackling complex political, cultural, and economic systems that are so deeply embedded within modern societal structures as to be virtually insurmountable. Under such structures, stark vulnerabilities exist and are nurtured, and the impacts of climatic extremes or environmental mismanagement are socially and economically biased. The richest members of society in the past built sturdy houses on prime land; they had savings and possibly even insurance; and much remains the same today, despite national housing and healthcare schemes in many parts of the world. The poorest did not, and still often do not, have access to risk mitigation strategies. Although our relationship with our climate has changed in modern times, this does not mean that we have become less dependent on, or influenced by, our climate, and climate change is creating a whole new set of extreme challenges that cities, nations, and individuals are now being forced to meet.

As a frame through which to view history, a climatic perspective can offer new ways of viewing the past, the present, and the future.[30] While this book does not focus on climate change in its Anthropogenic context, it hopes to bring into view a world where humans experienced a closer relationship with their weather and, in so doing, to bring our contemporary relationship with our climate into perspective, showing where fault lines exist. Given that climate, weather, and the impacts of their extremes reach across national borders and political structures, climate also has the power to bind disparate or far-flung communities in shared experiences. However, the experience of climate differs depending on who (and when) you are, and this is just as true now as it was in the past.

The focus here has been on a particular time and space, but more narrowly, the study has been based on documentation that represents British imperial interests and ways of thinking about the world through the lens of the weather. As Jürgen Renn notes in his mammoth study of knowledge, a history of science that has at its core specific case studies inevitably creates a fragmented picture.[31] Melding together a specific history of science with a localized and British history of the tropical environment, this book is thus limited in who, and what, it represents. This is not simply a product of the sources but the reality of science at the time. Despite the existence of meteorological officers in senior positions, such as Koruthu and Rajendram, it is my contention that, in the field of meteorology at least, the colonial authorities took on board only limited (or even no) information about the weather from the people indigenous to the Malayan Archipelago, nor did they employ non-Europeans in roles beyond basic levels until after World War Two. For the many Asian observers and technical staff who formed the backbone of the meteorological service, this was not a system of two-way knowledge exchange. The knowledge that impacted the service was from global sources—the changing understanding of local to regional and worldwide atmospheric forces that took place over the nineteenth century and continued in the new era of aviation, upper air, radar, and eventually, satellite observation and technologies. Likewise, in the many projects to improve water supply or to mitigate floods, the workers servicing these were low-skilled laborers or even prisoners, whose role it was to undertake the backbreaking manual labor in the tropical heat, while the plans and schemes came from imported European knowledge, firms, and companies. I would then suggest that we approach this subject with these limitations in mind and seek instead to apply the stories contained within these pages to a framework of understanding the history of weather science and controlling weather under colonialism, rather than trying to insert examples that were simply not there.

The history of colonial science cannot be viewed as a microcosm. It was influenced by global shifts and changing ideas in science, and the people involved in this story drew on wide networks of people and resources to develop their ideas. As the discussion of the desiccation debate or that of the evolution of meteorological services has demonstrated, local circumstances and global ideas integrated. The work that was undertaken to improve local observational coverage helped fill in gaps in tropical weather research for Southeast Asia, and the participation of Malayan directors and senior officers led to the colony having representation at meetings where the future of regional forecasting and collaboration was decided, a role that increased significantly over the twentieth century. In terms of controlling the weather and managing resources, this history of colonial successes or failures is a top-down but illuminating story of social inequality that can shed light on the uneven impacts of extreme weather and weather science on societies. As extreme events and uneven access to scientific education or scientific roles persist (and in some cases, worsen), these lessons have great relevance to our understanding of power dynamics in the bigger picture of tackling climate change.

NOTES

INTRODUCTION: THE DAWN OF IMPERIAL WEATHER

1. G. M. Reith, *Handbook to Singapore* (1892), revised by Walter Makepeace with an introduction by Paul Kratoska (Singapore: Oxford University Press, 1985), 125–27.

2. I am using the term "Malaya" to refer to the states of the Malay Peninsula, East Malaysia, and Singapore that came under British rule from 1786 to 1957.

3. Isabella Bird, letter 7, dated 19 January 1879, Singapore, in *The Golden Chersonese and the Way Thither* (New York: G. P. Putnam's Sons, 1883).

4. Walter Makepeace, Gilbert Brooke, and Roland St. John Braddell, eds., *One Hundred Years of Singapore* (London: John Murray, 1921), 1:492.

5. Captain Sherard Osborn, quoted in Makepeace, Brooke, and St. John Braddell, *One Hundred Years of Singapore*, 474.

6. Bird, *Golden Chersonese*, 81.

7. Ambrose B. Rathbone, *Camping and Tramping in Malaya: Fifteen Year Pioneering in the Native States of the Malay Peninsula* (London: Swan Sonnenschein and Co., 1898), 334–35.

8. Victor R. Savage, *Western Impressions of Nature and Landscape in Southeast Asia* (Singapore: NUS Press, 1984).

9. For a good overview of this literature, see Paul S. Sutter, "The Tropics: A Brief History of an Environmental Imaginary," in *The Oxford Handbook of Environmental History*, ed. Andrew C. Isenberg (Oxford: Oxford University Press, 2014), 178–204; David Arnold, *The Tropics and the Traveling Gaze: India, Landscape and Science, 1800–1856* (Seattle: University of Washington Press, 2006); Felix Driver and Luciana Martins, eds., *Tropical Visions in an Age of Empire* (Chicago: Chicago University Press, 2005).

10. Mark Harrison, *Climates and Constitutions: Health, Race, Environment and British Imperialism in India, 1600–1850* (New Delhi: Oxford University Press, 1999); Mark Harrison, *Public Health in British India: Anglo-Indian Preventative Medicine 1859–1914* (Cambridge: Cambridge University Press, 1994), 36.

11. David Arnold, *Colonizing the Body: State Medicine and Epidemic Disease in Nineteenth-Century India* (Berkeley: University of California Press, 1993), 37. See also Edward Said, Orientalism (New York, Pantheon Books, 1978); James Beattie, *Empire and Environmental Anxiety: Health, Science, Art and Conservation in South Asia and Australasia, 1800–1920* (Basingstoke: Palgrave Macmillan, 2011); D. Kennedy, "The Perils of the Midday Sun: Climatic Anxieties in the Colonial Tropics," in *Imperialism and the Natural World*, ed. J. M. MacKenzie (Manchester: Manchester University Press, 1990), 118–40.

12. Aitor Anduaga, *Politics, Statistics and Weather Forecasting, 1840–1910: Taming the Weather* (London: Routledge, 2020); Kristine C. Harper, *Weather by the Numbers: The Genesis of Modern Meteorology* (Cambridge, MA: MIT Press, 2012).

13. Exceptions include Robert M. Rouphail, "Cyclonic Ecology: Sugar, Cyclone Science, and the Limits of Empire in Mauritius and the Indian Ocean World, 1870s–1930s," *Isis* 110:1 (2019): 48–67; Martin Mahony, "The 'Genie of the Storm': Cyclonic Reasoning and the Spaces of Weather Observation in the Southern Indian Ocean, 1851–1925," *British Journal for the History of Science* 51:4 (2018): 607–33.

14. *First Report of the British Association for the Advancement of Science* (York: Thomas Wilson and Sons, 1832), 47–48.

15. Lee T. Macdonald, *Kew Observatory and the Evolution of Victorian Science, 1840–1910* (Pittsburgh, PA: University of Pittsburgh Press, 2018); Rebekah Higgitt, "A British National Observatory: The Building of the New Physical Observatory at Greenwich, 1889–1898," *British Journal for the History of Science* 47:4 (2013): 609–35; Mahony, "Genie of the Storm"; C. R. Carter, *Magnetic Fever: Global Imperialism and Empiricism in the Nineteenth Century* (Philadelphia, PA: American Philosophical Society, 2009); Dilip M. Salwi, "Madras Observatory: A Forgotten Page in Astronomy," *Journal of the British Astronomical Association* 98:4 (1988): 189–93.

16. Kevin MacKeown, *Early China Coast Meteorology: The Role of Hong Kong* (Hong Kong: Hong Kong University Press, 2011); J. Cawood, "The Magnetic Crusade: Science and Politics in Early Victorian Britain," *Isis* 70:4 (1979): 492–518.

17. Katherine Anderson, "Looking at the Sky: The Visual Context of Victorian Meteorology," *British Journal for the History of Science* 36:3 (2003): 305–6; Simon Naylor, "Log Books and the Law of Storms: Maritime Meteorology and the British Admiralty in the Nineteenth Century," *Isis* 106:4 (2015): 771–97; Simon Naylor, "Nationalising Provincial Weather: Meteorology in Nineteenth-Century Cornwall," *British Journal for the History of Science* 39:3 (2006): 407–33.

18. Sarah Carson, "Anticipating the Monsoon: The Necessity and Impossibility of the Seasonal Weather Forecast for South Asia, 1886–1953," *British Journal for the History of Science* 54:3 (2021): 305–25; Giuditta Parolini, "Building Networks of Exchange in Agricultural Meteorology: The Agro-meteorological Service in French Indochina," *History of Meteorology* 9 (2020): 1–20.

19. Correspondence from The High Commissioner FMS to L. C. M. S. Amery Colonial Office and Memorandum on a pamphlet entitled "A Meteorological Department for Malaya," by Sir George Maxwell and Herbert C. Robinson, Colonial Office (hereafter CO) 273/541, The National Archives UK (hereafter TNA); John K. Corner, *My Father in His Suitcase: In Search of E. J. H. Corner, the Relentless Botanist* (Singapore: Landmark Books, 2013), 84, 86, 179; Timothy P. Barnard and Joanna W. C. Lee, "A Spiteful Campaign: Agriculture, Forests, and Administering the Environment in Imperial Singapore," *Environmental History* 27:3 (2022): 479–80.

20. Andrew Goss, "Introduction: An Imperial Turn in the History of Science," in *The Routledge Handbook of Science and Empire*, ed. Andrew Goss (London: Routledge, 2021), 1–9; Simon Naylor and Matthew Goodman, "Atmospheric Empire: Historical Geographies of Meteorology at the Colonial Observatories," in *Weather, Climate, and the Geographical Imagination*, ed. Martin Mahony and Samuel Randalls (Pittsburgh: University of Pittsburgh Press, 2020), 25–42; Martin Mahony, "For an Empire of 'All Types of Climate': Meteorology as an Imperial Science," *Journal of Historical Geography* 51 (2016): 29–39; Roy Macleod, "Passages in Imperial Science: From Empire to Commonwealth," *Journal of World History* 4:1 (1993): 117–50.

21. For way markers on this historiographic journey, see Joseph M. Hodge, *Triumph of the Expert: Agrarian Doctrines of Development and the Legacies of British Colonialism* (Athens: Ohio University Press, 2007); Alan Lester, "Imperial Circuits and Networks: Geographies of the British Empire," *History Compass* 4 (2006): 124–41; K. Raj, *Relocating Modern Science: Circulation and the Construction in South Asia and Europe, 1650–1900* (Basingstoke: Palgrave Macmillan 2007); James A. Secord, "Knowledge in Transit," *Isis* 95 (2004): 654–72; David N. Livingstone, *Putting Science in Its Place: Geographies of Scientific Knowledge* (Chicago: Chicago University Press, 2004); Bernard Cohn, *Colonialism and Its Forms of Knowledge: The British in India* (Princeton, NJ: Princeton University Press, 1996); Adi Ophir and Stephen Shapin, "The Place of Knowledge: A Methodological Survey," *Science in Context* 4 (1991): 3–21.

22. Jürgen Renn, *The Evolution of Knowledge: Rethinking Science for the Anthropocene* (Princeton, NJ: Princeton University Press, 2020), 14; Martin Mahony and Samuel Randalls, *Weather, Climate, and the Geographical Imagination*, 7–8; Deborah Coen, *Climate in Motion: Science, Empire, and the Problem of Scale* (Chicago: University of Chicago Press, 2018); Lester, "Imperial Circuits and Networks," 132; Christian Jennings, "Unexploited Assets: Imperial Imagination,

Practical Limitations, and Marine Fisheries Research in East Africa, 1917–53," in *Science and Empire: Knowledge and Networks of Science across the British Empire, 1800–1970*, ed. Brett M. Bennett and Joseph M. Hodge (London: Palgrave Macmillan, 2011), 253–74.

23. Emily O'Gorman, "'Soothsaying' or 'Science'? H. C. Russell, Meteorology, and Environmental Knowledge of Rivers in Colonial Australia," in *Climate, Science, and Colonization: Histories from Australia and New Zealand*, ed. James Beattie, Emily O'Gorman, and Matthew Henry (New York: Palgrave Macmillan, 2014), 181.

24. Mark E. Frank, "Frontier Atmosphere: Observation and Regret at Chinese Weather Stations in Tibet, 1939–1949," *British Journal for the History of Science* 54:3 (2021): 361–79.

25. Mahony, "For an Empire," 171. See also Suman Seth, "Putting Knowledge in Its Place: Science, Colonialism, and the Postcolonial," *Postcolonial Studies* 12:4 (2009): 380.

26. Recent work in this area includes Harriet Mercer, "Atmospheric Archives: Gender and Climate Knowledge in Colonial Tasmania," *Environment and History* 27:2 (2021): 193–210; Frank, "Frontier Atmosphere"; Ruth A. Morgan, "Prophecy and Prediction: Forecasting Drought and Famine in British India and the Australian Colonies," *Global Environment* 13 (2020): 96–133.

27. David Pretel, "Hidden Connections: The Global History of Jungle Commodities," *Technology and Culture* 64:1 (2023): 202–19; Londa Schiebinger, *Plants and Empire: Colonial Bioprospecting in the Atlantic World* (Cambridge, MA.: Harvard University Press, 2004); Coen, *Climate in Motion*; Philipp Lehmann, *Desert Edens: Colonial Climate Engineering in an Age of Anxiety* (Princeton, NJ: Princeton University Press, 2022); Thomas Simpson, "Imperial Slippages: Encountering and Understanding Ice in Colonial India," in *Ice Humanities: Living, Thinking, and Working in a Melting World*, ed. Klaus Dodds and Sverka Sörlin (Manchester: Manchester University Press, 2022).

28. For example Geoffrey K. Pakiam, "'Not the Oil of the Country': Smallholders and British Malaya's Oil Palm Industry, 1929–1941," *Agricultural History* 95:1 (2021): 69–103; Lynn Hollen Lees, *Planting Empire, Cultivating Subjects: British Malaya, 1786–1941* (Cambridge: Cambridge University Press, 2017); Jeyamalar Kathirithamby-Wells, "The Implications of Plantation Agriculture for Biodiversity in Peninsular Malaysia: A Historical Analysis," in *Beyond the Sacred Forest: Complicating Conservation in Southeast Asia*, ed. Michael R. Dove, Percy E. Sajise, and Amity A. Doolittle (Durham, NC: Duke University Press, 2011), 62–90; Jeyamalar Kathirithamby-Wells, *Nature and Nation: Forests and Development in Peninsular Malaysia* (Singapore: NUS Press, 2005). Richard Grove pioneered the study of colonial nature and weather science in two books: Richard Grove, *Ecology, Climate, and Empire: Colonialism and Global Environmental History, 1400–1940* (Cambridge: Cambridge University Press, 1997), and Richard Grove, *Green Im-*

perialism: Colonialism, Tropical Island Edens and the Origins of Environmentalism, 1600–1860 (Cambridge: Cambridge University Press, 1995).

29. Makepeace, Brooke, and St John Braddell, *One Hundred Years of Singapore*, 2:166.

30. Richard T. Corlett, "The Ecological Transformation of Singapore, 1819–1990," *Journal of Biogeography* 19:4 (1992): 411–20.

31. Wazir Jahan Karim, "The 'Discovery' of Penang Island at Tanjong Tokong before 1785: Bapu Alaidin Meera Hussein Lebai and Captain Francis Light," *Journal of the Malaysian Branch of the Royal Asiatic Society* (hereafter *JMBRAS*) 86:1 (2013): 1–29; R. Bonney, "Francis Light and Penang," *JMBRAS* 38:1 (1965): 135–58.

32. The post of Resident is an administrative post, loosely based on the Indian residency system. The Resident acts as an advisor to the Malay Susan of a particular area but has the de facto right to rule in all administrative matters.

33. Mike Davis, *Late Victorian Holocausts: El Niño Famines and the Making of the Third World* (London: Verso, 2001).

34. Ramachandra Guha, *The Unquiet Woods: Ecological Change and Peasant Resistance in the Himalaya* (Berkeley: University of California Press, 2000).

35. Richard Grove and J. Chappell, *El Niño: History and Crisis: Studies from the Asia-Pacific Region* (Cambridge: White Horse Press, 2000); Grove, *Ecology, Climate, and Empire*; Jan Golinski, *Making Natural Knowledge: Constructivism and the History of Science* (Cambridge: Cambridge University Press, 1998); Andrew Pickering, "From Science as Knowledge to Science as Practice," in *Science as Practice and Culture*, ed. Andrew Pickering (Chicago: Chicago University Press, 1992), 1–26; Steven Shapin and Simon Schaffer, *Leviathan and the Air-Pump: Hobbes, Boyle, and the Experimental Life* (Princeton, NJ: Princeton University Press, 1985).

36. Robert E. Kohler, *Landscapes and Labscapes: Exploring the Lab-Field Border in Biology* (Chicago: Chicago University Press, 2002); Alex Soojung-Kim Pang, *Empire and the Sun: Victorian Solar Eclipse Expeditions* (Stanford, CA: Stanford University Press, 2002); Oliver Hochadel and Agusti Nieto-Galen, eds., *Urban Histories of Science: Making Knowledge in the City, 1820*–1940 (London: Routledge, 2018).

37. Frank, "Frontier Atmosphere"; Anya Zilberstein, *A Temperate Empire: Making Climate Change in Early America* (New York: Oxford University Press, 2016); Gregory T. Cushman, "Humboldtian Science, Creole Meteorology, and the Discovery of Human-Caused Climate Change in South America," *Osiris* 26 (2011): 19–44; Masumi Zaiki and Togo Tsukahara, "Meteorology on the Southern Frontier of Japan's Empire: Ogasawara Kazuo at Taihoku Imperial University," *East Asian Science, Technology and Society: An International Journal* 1 (2007): 183–203.

38. For example, David Brewster, "On the Mean Temperature of the Equator, as deduced from Observations made at Prince of Wales Island, Singapore and Malacca," *Edinburgh Journal of Science* 8 (1828): 60–67.

39. For example, see James Horsburgh, *The India Directory or Directions for Sailing to and from the East Indies, China, Australia and the Interjacent Ports of Africa and South America Vol. II* (London, 1843); Henry Piddington, "Fifteenth Memoir of the Law of Storms," *Journal of the Asiatic Society of Bengal* 17 (1848): 27–56; Charles Meldrum, *Contributions to the Meteorology and Hydrography of the Indian Ocean: a Meteorological Journal of the Indian Ocean for the month of March 1853* (1854); M. F. Maury, *Explanations and Sailing Directions to Accompany the Wind and Current Charts* (1858); F. Labrosse, *The Navigation of the Pacific Ocean and China Seas*, trans. J. W. Miller (1875); W. H. Rosser, *The Law of Storms Considered Practically* (1876).

40. Coen, *Climate in Motion*.

41. For example, William Farquhar, "Thermometrical and Barometrical Tables communicated by Lieut. Col. William Farquhar," *Transactions of the Royal Asiatic Society of Great Britain and Ireland* 1 (1827): appendix 1, 585–99; Joseph S. Travelli, "Abstract of a Meteorological Register kept at Singapore," *American Journal of Science and Arts* 44 (1843): 150–58.

42. Lehmann, *Desert Edens;* Stephen Legg, "Debating the Climatological Role of Forests in Australia, 1827–1949: A Survey of the Popular Press," in James Beattie, Emily O'Gorman, and Matthew Henry, eds., *Climate, Science, and Colonization: Histories from Australia and New Zealand* (New York: Palgrave Macmillan, 2014), 119–36. For contemporary studies on how urbanization was affecting temperature, see Luke Howard, *The Climate of London* (London, 1833), vols. 1–3, and Emile Renou, "Temperatures Differences between City and Country," *Annuales du Bureau central météorologique de France*, 3:79 (1855), revised 1868.

43. Henry F. Blanford, "On the barometric see-saw between Russia and India in the sunspot cycle," *Nature* 21 (1880), 477; Charles Todd (with H. C. Russel and R. L. J. Ellery), "The Meteorologist: Droughts in Australia," *The Australasian*, 29 December 1888, 40. See also Morgan, "Prophecy and Prediction"; Richard Grove and George Adamson, *El Niño in World History* (Basingstoke: Palgrave, 2017).

44. MacKeown, *Early China Coast Meteorology*.

45. Brenda Yeoh, "Urban Sanitation, Health and Water Supply in Late Nineteenth and Early Twentieth Century Colonial Singapore," *Southeast Asia Research* 1:2 (1993): 143–72.

46. William Beinart and Lotte Hughes, *Environment and Empire* (Oxford: Oxford University Press, 2007), ch. 8; Satyajit Singh, *Taming the Waters: The Political Economy of Large Dams in India* (Delhi: Oxford University Press, 1997); David Arnold and Ramachandra Guha, eds., *Nature, Culture, Imperialism: Essays on the Environmental History of South Asia* (Oxford: Oxford University Press, 1996).

47. Karl Wittfogel, *Oriental Despotism: A Comparative Study of Total Power* (New Haven: Yale University Press, 1957).

48. Kae Takarabe, "The Smithsonian Meteorological Project and Hokkaido, Japan," *History of Meteorology* 9 (2020): 1–23; Parolini, "Building Networks."

49. Mike Hulme, "Climate and Its Changes: A Cultural Appraisal," *Geography and Environment* 2:1 (2015): 2.

50. See, for example, Heinrich Barth's or Mungo Park's writing on the African interior; J. R. Logan, "The Probable Effects on the Climate of Pinang of the Continued Destruction of its Hill Jungles," *Journal of the Indian Archipelago and Eastern Asia* 2 (1848): 534–36.

51. Gregory A. Barton, *Empire Forestry and the Origins of Environmentalism* (Cambridge: Cambridge University Press, 2002).

52. Hugh Cleghorn, *The Forests and Gardens of South India* (London: W. H. Allen, 1861), v–vi.

53. Pretel, "Hidden Connections," 204.

54. Nathaniel Cantley, "Report on the Forests of the Straits Settlements, 1882," *Annual Report on the Botanic Gardens, Singapore, 1884* (Singapore, 1885), 1–7.

55. Fiona Williamson, "Local Problems, Global Solutions? Making It Rain in Hong Kong c. 1890–1930," *History of Science*, early online version published April 2023; Lehmann, *Desert Edens*; Meredith McKittrick, "Talking about the Weather: Settler Vernaculars and Climate Anxieties in Early Twentieth-Century South Africa," *Environmental History* 23 (2018): 3–27; Meredith McKittrick, "Theories of 'Reprecipitation' and Climate Change in the Settler Colonial World," *History of Meteorology* 8 (2017): 74–94.

56. G. Alphonso and Albert Lau, *The Causeway* (Kuala Lumpur and Singapore: National Archives of Malaysia and National Archives of Singapore, 2011).

57. Kathirithamby-Wells, *Nature and Nation*, 43, 59.

58. Sheila Jasanoff, "Future Imperfect: Science, Technology, and the Imaginations of Modernity," in *Sociotechnical Imaginaries and the Fabrication of Power*, ed. Sheila Jasanoff and Sang-Hyun Kim (Chicago: University of Chicago Press, 2015), 19.

PART I: IMPERIAL WEATHER

1. *Special Report on Weather Forecasting in Malaya by C. D. Stewart, 1935*, 1, SEL: SEC G.438–95, National Archives of Malaysia (NAM).

2. Robert Marc Freidman, *Appropriating the Weather: Vilhelm Bjerknes and the Construction of a Modern Meteorology* (Ithaca: Cornell University Press, 1989).

3. Kristine C. Harper, *Weather by the Numbers: The Genesis of Modern Meteorology* (Cambridge, MA: MIT Press, 2012), 49–68.

4. *Special Report on Weather Forecasting*, 8.

CHAPTER 1: ATMOSPHERIC STATES

1. Aitor Anduaga, *Politics, Statistics and Weather Forecasting, 1840–1910: Taming the Weather* (London: Routledge, 2020), 12; Kerby Alvarez, "Instrumentation

and Institutionalization: Colonial Science and the Observatorio Meteorológico de Manila, 1865–1899," *Philippine Studies: Historical and Ethnographic Viewpoints* 63:3–4 (2016): 385–416.

2. Greg Bankoff, "Aeolian Empires: The Influence of Winds and Currents on European Maritime Expansion in the Days of Sail," *Environment and History* 23:2 (2017): 163–96; James F. Warren, "Weather, History and Empire: The Typhoon Factor and the Manila Galleon Trade, 1565–1815," in *Anthony Reid and the Study of the Southeast Asian Past*, ed. Geoff Wade and Tana Li (Singapore: Institute of Southeast Asian Studies, 2012), 183–220.

3. Thomas Simpson, "Imperial Slippages: Encountering and Understanding Ice in Colonial India," in *Ice Humanities: Living, Thinking and Working in a Melting World*, ed. Klaus Dodds and Sverker Sörlin (Manchester: Manchester University Press, 2022), 205–27; Deborah Coen, *Climate in Motion; Science, Empire, and the Problem of Scale* (Chicago: Chicago University Press, 2018).

4. Katherine Anderson, "Mapping Meteorology," in *Intimate Universality: Local and Global Themes in the History of Weather and Climate*, ed. James R. Fleming, Vladimir Jankovic, and Deborah R. Coen (Sagamore Beach, MA: Watson Publishing International, 2006), 69–91; Katherine Anderson, *Predicting the Weather: Victorians and the Science of Meteorology* (Chicago: Chicago University Press, 2005); Mark Monmonier, *Air Apparent: How Meteorologists Learned to Map, Predict, and Dramatize Weather* (Chicago: University of Chicago Press, 1999).

5. Mark Frank, "Frontier Atmosphere: Observation and Regret at Chinese Weather Stations in Tibet, 1939–1949," *British Journal for the History of Science* 54:3 (2021): 361–79; Simon Naylor, "Log-Books and the Law of Storms: Maritime Meteorology and the British Admiralty in the Nineteenth Century," *Isis* 106:4 (2015): 771–97; Andrea Westermann, "Disciplining the Earth: Earthquake Observation in Switzerland and Germany at the Turn of the Nineteenth Century," *Environment and History* 17 (2011): 53–77; David Aubin, Charlotte Bigg, and Otto H. Sibum, eds., *Observatories and Astronomy in Nineteenth-Century Science and Culture* (Durham, NC: Duke University Press, 2010); Alex Soojung-Kim Pang, *Empire and the Sun: Victorian Solar Eclipse Expeditions* (Stanford, CA: Stanford University Press, 2002); John Cawood, "The Magnetic Crusade: Science and Politics in Early Victorian Britain," *Isis* 70:4 (1979): 492–518.

6. Anduaga, *Politics, Statistics and Weather Forecasting*; Azadeh Achbari and Frans van Lunteren, "Dutch Skies, Global Laws: The British Creation of Buys Ballot's Law," *Historical Studies in the Natural Sciences* 46:1 (2016): 1–43.

7. Christa Jungnickel and Russell McCormmach, *Intellectual Mastery of Nature: Theoretical Physics from Ohm to Einstein. The Torch of Mathematics, 1800–1870* (Chicago: Chicago University Press, 1990), vol. 1; Cawood, "Magnetic Crusade."

8. Richard Hamblyn, *The Invention of Clouds: How an Amateur Meteorologist Forged the Language of the Skies* (London: Picador, 2011), esp. 90–111.

9. Aitor Anduaga, *Politics, Statistics and Weather Forecasting, 1840–1910* (London: Routledge, 2020), 106–200: Peter Moore, *The Weather Experiment: The Pioneers Who Sought to See the Future* (London: Penguin Random House, 2015), esp. 97–166; James R. Fleming, *Inventing Atmospheric Science: Bjerknes, Rossby, Wexler, and the Foundations of Modern Meteorology* (Cambridge, MA: MIT Press, 2016), 13–76.

10. Mahani Musa, "Early History: Penang before 1786," in *Early History of Penang*, ed. Muhammad Haji Salleh (Penang: Penerbit USM, 2012), 30; R. Bonney, "Francis Light and Penang," *JMBRAS* 38:1 (1965): 142; D. K. Bassett, "British Trade and Policy in Indonesia, 1760–1772," *Bijdragen tot de Taal, Land en Volkenkunde*, 120, 2de Afl. (1964): 197–223.

11. Norman Macalister, *Historical Memoir Relative to Prince of Wales Island* (London: J. H. Hart, 1803), 8.

12. N. Wallich, Superintendent Botanic Garden [Calcutta] To C. Lushington, Secretary to the Government in the General Department, 7 January 1823, copy, Bengal Public Consultations: Fort William 23rd January 1823, No. 45, India Office Records (IOR)/P/10/61, British Library (BL).

13. J. S. M. Coleman and K. T. Law, "Meteorology," *Reference Module in Earth Systems and Environmental Sciences: Science Direct* (2015): 6.

14. M. Chenoweth, J. M. Vaquero, R. Garcia-Herrera, and D. Wheeler, "A Pioneer in Tropical Meteorology: William Sharpe's Barbados Weather Journal, April–August 1680," *Bulletin of the American Meteorological Society* 88:12 (2007): 1958.

15. Chenoweth, Vaquero, Garcia-Herrera, and Wheeler, "Pioneer in Tropical Meteorology"; Jan Golinski, "Barometers of Change: Meteorological Instruments as Machines of Enlightenment," in *The Sciences in Enlightened Europe*, ed. William Clark, Jan Golinski, and Simon Schaffer (Chicago: University of Chicago Press, 1999), 73; Captain Joseph Huddart, *The Oriental Navigator* (1801).

16. Edmond Halley, "An Historical Account of the Trade Winds and monsoons, observable in the Tropicks, with an attempt to assign the physical cause of the said Winds," *Philosophical Transactions* 16 (1686): 153–68; Henry F. Blanford, "On the Origin of a Cyclone," *Proceedings of the Royal Society* 17 (1869): 472–82; Ralph Abercromby, "On the relation between tropical and extra-tropical cyclones," *Proceedings of the Royal Society* 43 (1888): 1–30. For the history, see G. Kutzbach, *The Thermal Theory of Cyclones: A History of Meteorological Thought in the Nineteenth Century* (Boston, MA: American Meteorological Society, 1979).

17. James Horsburgh, *The India Directory of Directions for Sailing to and from the East Indies, China, Australia and the Interjacent Ports of Africa and South America* (London, 1843), 1:289.

18. Sarah Carson, "Anticipating the Monsoon: The Necessity and Impossibility of the Seasonal Weather Forecast for South Asia, 1886–1953," *British Journal for the History of Science* 54:3 (2021): 305–25; Chi Chi Huang, "'Hong Kong Can

Afford a Typhoon or Two': British Discussions of Revolving Storms," *British Journal for the History of Science* (2021): 327–39.

19. Lawrence Rotch, "The Organisation of the Meteorological Service in some of the Principal countries of Europe," *American Meteorological Journal* 5:9 (1889): 393.

20. Jim Burton, "Robert Fitzroy and the Early History of the Meteorological Office," *British Journal for the History of Science* 19:2 (1986): 147–76.

21. For example, see Clements R. Markham, ed., *The Voyages of Sir James Lancaster, Knight, to the East Indies &c.* (London: Hakluyt Society, 1877); Alexander Hamilton, *A New Account of the East Indies, from the year 1688–1723* (Edinburgh: John Mosman, 1727), vols. 1–2.

22. Charles Burton Buckley, *An Anecdotal History of Singapore 1819–1867* (Oxford University Press: Oxford, 1984), 736; *The Singapore Free Press and Mercantile Advertiser*, 15 October 1835, 1.

23. For biographical detail on Low, see Papers of James Low, catalogued by Nancy Charley, Royal Asiatic Society Archives, at https://archiveshub.jisc.ac.uk/search/archives/eeab4301-e9fc-3598-8826-5bb5fd1c4bb7?terms=%22Low%20James%201791-1852%22/.

24. *Singapore Free Press and Mercantile Advertiser*, 27 October 1836, 2; James Low, *A Dissertation on the Soil & Agriculture of the British Settlement of Penang, Or Prince of Wales Island, in the Straits of Malacca: Including Province Wellesley on the Malayan Peninsula. With Brief References to the Settlements of Singapore & Malacca* (Singapore: Singapore Free Press Office, 1836), 817–18.

25. T. M. Ward and J. P. Grant, "Contributions to the Medical Topography of Prince of Wales Island or Pulo Pinang," in *Official Papers on the Medical Statistics and Topography of Malacca and Prince of Wales' Island . . . &c* (Pinang: Government Press, 1830), 50.

26. J. Turnbull Thomson, "A Sketch of the Career of the Late James Richardson Logan of Penang and Singapore," *Journal of the Straits Branch of the Royal Asiatic Society* 7 (1881): 78.

27. J. R. Logan, "Sketch of the physical geography and geology of the Malay Peninsula," *Journal of the Indian Archipelago and Eastern Asia* 2 (1848), 114.

28. Logan, "Sketch of the physical geography and geology of the Malay Peninsula," 108.

29. Cawood, "Magnetic Crusade."

30. Cawood, "Magnetic Crusade," 492.

31. Matthew Goodman, "Follow the Data: Administering Science at Edward Sabine's Magnetic Department, Woolwich, 1841–57," *Notes and Records: The Royal Society Journal of the History of Science* 73:2 (2019): 192.

32. J. R. Logan, ed., "Miscellaneous Notices &c: The Singapore Observatory," *Journal of the Indian Archipelago and Eastern Asia* 3 (1849): xxxviii–xlvi.

33. C. M. Elliot, "Singapore Observatory: Remarks on the manner in which

the Stations adjacent to Singapore might be surveyed, with little, if any expense to Government" (1844), MM/10/173, Royal Society (hereafter RS).

34. Letter dated 7 January 1843, IOR/F/4/2026, 91387, f. 12, BL. See also Edward Belcher, *Narrative of the Voyage of the HMS Samarang, During the years 1843–46: Employed Surveying the Islands of the Eastern Archipelago* (repr.; Cambridge, Cambridge University Press, 2011), 166.

35. C. M. Elliot, "Magnetic Survey of the Eastern Archipelago," *Abstracts of the Papers Communicated to the Royal Society, Vol. 6*, 1850–54, 15, at www.jstor.org/stable/111125/; *Report of the Fifteenth Meeting of the British Association for the Advancement of Science, June 1845* (London: J. Murray, 1846), 68.

36. *The Straits Times*, 31 December 1845, 2.

37. The Singapore Observatory Magnetic Yearbooks have been digitized and made freely available online by the British Geological Survey, at http://www.geomag.bgs.ac.uk/data_service/data/yearbooks/sin.html/.

38. William Jacob to Thomas Maclear, 21 January 1850, fol. 208, MS 15.57, Royal Greenwich Observatory (RGO), Cambridge University Library (CUL).

39. Robert Little, "Part I: On the medical Topography of Singapore, particularly its Marshes and Malaria," *Journal of the Indian Archipelago and Eastern Asia* 5 (1851): 452.

40. Little, "Part I: On the medical Topography."; Anon. [Elliot], "Miscellaneous Notices &c: The Singapore Observatory," *Journal of the Indian Archipelago and Eastern Asia* 3:6 (1849): xxxix.

41. Anon., "Miscellaneous Notices," xliii.

42. Anon., "Miscellaneous Notices," xliv.

43. Robert Little, "An Essay on Coral Reefs as the Cause of Blakan Mati Fever," *Journal of the Indian Archipelago and Eastern Asia* 3 (1849): 413–44; Anon., "Remarks on Dr Little's 'Essay on Coral Reefs as the cause of Fever'," *Journal of the Indian Archipelago and Eastern Asia* 3 (1849): 701.

44. Anon., "Miscellaneous Notices," xliv.

45. Logan's editorial note in Anon., "Miscellaneous Notices," xliv. See also Thomson, "Sketch of the Career of the Late James Richardson Logan," 77.

46. James D. Forbes, "Report on the Recent Progress and Present State of Meteorology," in British Association for the Advancement of Science (BAAS), *Report of the First and Second Meetings 1831, 1832: Second Report* (London: John Murray, 1833): 196, 199.

47. See, for example, L. F. Kämtz, *A Complete Course of Meteorology*, trans. C. V. Walker (London: Hippolyte Baillière, 1845); Heinrich W. Dove and Ludwig Moser, *Repertorium der Physik* (Berlin: Verlag von Veit & Comp., 1837).

48. BAAS, "Recommendations of the Sub-Committees: Meteorology," *Report of the Second Meeting* (1834): 49. John Goldingham's work at Madras as first British Madras observatory astronomer (1802) was a notable exception. See J. Goldingham, *Madras Observatory Papers* (Madras: College Press of Fort St. George,

1827). See also BAAS, "Recommendations of the Sub-Committees: Committee of Mathematical and Physical Science," *Proceedings of the General Meeting, 1831* (1831): 47–51, esp. 48.

49. Forbes, "Supplementary Report on Meteorology," 52, 55.

50. Anduaga, *Politics, Statistics and Weather Forecasting*, 79–80. Heinrich W. Dove, "Temperature Tables, with Introductory Remarks by Edward Sabine" *Report of the Seventeenth Meeting of the British Association for the Advancement of Science, Held at Oxford 1847* (London: John Murray, 1848), 373–465.

51. John F. Daniell, *Elements of Meteorology*, 3rd ed. (London: John W. Parker, 1845), vol. 1, esp. 133–39, 153–54. For example, both Forbes's and Humboldt's work is cited in Logan's *Journal of the Indian Archipelago and Eastern Asia* (1849), 3: xlv, 8, 122–23, 244, 449.

52. William Reid to John Herschel, 3 January 1839, DM/3/117–18, RS.

53. Extracts from Observations Book, 1827, Tavoy, Royal Commonwealth Society (RCS) 65/4/15, CUL. The Singapore observations were printed in the *Singapore Free Press and Mercantile Advertiser*, but there is not one complete year.

54. For example, "Meteorological Register of the Raffles Lighthouse, for the month of December 1864." *Straits Government Gazette* (Singapore, 1864): 86.

55. Heinrich W. Dove, "Temperature Tables, with Introductory Remarks by Edward Sabine," *Report of the Seventeenth Meeting of the British Association for the Advancement of Science, Held at Oxford 1847* (London: John Murray, 1848), 373–465; "Meteorological Register of the Raffles Lighthouse," table 16, India, China, and the Malay Archipelago.

56. *Straits Settlements Government Gazette* (hereafter *SSGG*), 13 August 1875, 557.

57. Timothy P. Barnard, *Nature's Colony: Empire, Nation and Environment in the Singapore Botanic Gardens* (Singapore: NUS Press, 2016), 20.

58. Resident Selangor, 1887, KL629/87, 1886, NAM.

59. C. M. Turnbull, *The Straits Settlements, 1826–67: Indian Presidency to Crown Colony* (London: Athlone Press, 1972), 143.

60. "Delta," untitled article, *Straits Times*, 9 June 1877, 28.

61. "Letter from the Governor of Prince of Wales Island, Singapore and Malacca to J. P. Grant, Secretary to the Government of Bengal, conveying the arguments which may be used for and against the clearance of the forest lands by the Grand Jury, 26 Aug 1851–24 Oct 1851," Home (Public) Department, Consultation 24 October 1851, No. 12, National Archives of India; "Forests, Prince of Wales's Island, clearance of left to discretion of local government, 1853–1854," IOR/Z/E/4/24/F329, BL.

62. J. R. Logan, "The Probable Effects on the Climate of Pinang of the Continued Destruction of its Hill Jungles," *Journal of the Indian Archipelago and Eastern Asia* 2 (1848): 534–36

63. Lenore Manderson, "Health Services and the Legitimation of the Colonial

State: British Malaya, 1786–1941," *International Journal of Health Services* 17:1 (1987): 91–112, esp. 101.

64. David Arnold, *The Tropics and the Traveling Gaze: India, Landscape and Science, 1800–1856* (Seattle: University of Washington Press, 2006); Mark Harrison, *Public Health in British India: Anglo-Indian Preventative Medicine, 1859–1914* (Cambridge: Cambridge University Press, 1994); David Arnold, *Colonizing the Body: State Medicine and Epidemic Disease in Nineteenth Century India* (Berkeley: University of California Press, 1993). See also James Johnson, *The Influence of Tropical Climates, More Especially of the Climate of India, on European Constitutions &etc.* (1815); J. R. Martin, *The Influence of Tropical Climates on European Constitutions &c* (London, 1853; new edn., 1856); R. Scoresby Jackson, *Medical Climatology* (London, 1862).

65. Noman Lockyer and W. W. Hunter, "Sunspots and famines," *Nineteenth Century* 2 (1877): 583–602; Anderson, *Predicting the Weather*, 274–75.

66. For example, J. L. Bryden, *Epidemic cholera in the Bengal Presidency: A Report on the cholera of 1866–68, and its relations to the cholera of previous epidemics* (Calcutta: Office of the Superintendent of Government Printing, 1869); J. M. Cunningham, *Report on the cholera epidemic of 1875 in India* (Calcutta: Office of the Superintendent of Government Printing, 1876), 6.

67. For more on the 1877–1878 event, see Don Garden, *Droughts, Floods & Cyclones: El Niños That Shaped Our Colonial Past* (Melbourne: Australian Scholarly Publishing, 2009); Mike Davis, *Late Victorian Holocausts: El Niño Famines and the Making of the Third World* (Verso, London: 2001); Henry Blanford, "The Eleven-Year Periodical Fluctuation of the Carnatic Rainfall," *Nature* 36 (1887): 227–29.

68. For more on the history of cholera in Singapore, see Loh Kah Seng and Li Yan Hsu, *Pandemics in Singapore, 1819–2022: Lessons for the Age of Covid-19* (Abingdon: Routledge, 2024), 94–114.

69. Hampshire quoted by Robert L. Jarman, ed., *Annual Reports of the Straits Settlements, 1855–1941*, 12 vols. (Slough: Archive Editions, 1998), 2:340.

70. Nobuhiko Endo and Jun Matsumoto, "A Brief History of Meteorological Observations in French Indochina," *Geographical Reports of Tokyo Metropolitan* 54 (2019): 36.

71. Aitor Anduaga, *Cyclones and Earthquakes: The Jesuits, Prediction, Trade, and Spanish Dominion in Cuba and the Philippines, 1850–1898* (Manila: Ateneo de Manila University Press, 2017); Kevin MacKeown, *Early China Coast Meteorology: The Role of Hong Kong* (Hong Kong: Hong Kong University Press, 2011)., ch. 1.

72. Huib J. Zuidervaart and Rob H. Van Gent, "'A Bare Outpost of Learned European Culture on the Edge of the Jungles of Java': Johan Maurits Mohr (1716–1775) and the Emergence of Instrumental and Institutional Science in Dutch Colonial Indonesia," *Isis* 95 (2004): 32–33.

73. Peter Boomgaard, "Introduction. From the Mundane to the Sublime: Science, Empire, and the Enlightenment (1760s–1820s)," in *Empire and Science in the Making: Dutch Colonial Scholarship in Comparative Perspective, 1760–1830*, ed. Peter Boomgaard (New York: Palgrave Macmillan, 2013); Andrew Goss, *The Floracrats: State-Sponsored Science and the Failure of the Enlightenment in Indonesia* (Madison: University of Wisconsin Press, 2011).

74. "Survey of Mount Faber Range and country surrounding Singapore, 1882," media image no. D2014040098, C.O. 129/205, National Archives of Singapore (NAS).

75. Walter Makepeace, Gilbert E. Brooke, Roland St. John Braddell, eds., *One Hundred Years of Singapore*, 2 vols. (London: John Murray, 1921), 1:481; *SSGG*, 9 May 1884, 532.

76. "Correspondence regarding the time ball for the Straits Settlements (Singapore) on Mount Serapong, 1888–1892," Papers of Frank Dyson, RGO 8/147, CUL.

77. Helen Tilley, *Africa as a Living Laboratory: Empire, Development, and the Problem of Scientific Knowledge, 1870–1950* (Chicago: Chicago University Press, 2011), 10.

78. At this point in time, published weather statistics from the Medical Department disappear from the newspapers and gazettes.

CHAPTER 2: MAKING METEOROLOGY MAINSTREAM

1. For more, see Matthew Henry, "Assembling the Weather: Expertise, Authority and the Negotiation of Trans-Tasman Aviation Forecasts," *History of Meteorology* 8 (2017): 179–201; James R. Fleming, *Inventing Atmospheric Science: Bjerknes, Rossby, Wexler, and the Foundations of Modern Meteorology* (Cambridge, MA.: MIT Press, 2016); Chihyung Jeon, "Flying Weather Men and Robot Observers: Instruments, Inscriptions, and Identities in US Upper-Air Observation, 1920–1940," *History and Technology* 26:2 (2010): 119–45; Gregory Cushman, "The Struggle over the Airways in the Americas, 1919–1945: Atmospheric Science, Aviation Technology, and Neocolonialism," in James R. Fleming, Vladimir Jankovic, and Deborah Coen, eds., *Intimate Universality: Local and Global Themes in the History of Weather and Climate* (Sagamore Beach, MA: Science History Publications: 2006), 175–222.

2. Roger Turner, "Weathering Heights: The Emergence of Aeronautical Meteorology as an Infrastructural Science," publicly available Penn Dissertations, University of Pennsylvania, 2010.

3. Robert Marc Friedman, *Appropriating the Weather: Vilhelm Bjerknes and the Construction of a Modern Meteorology* (Ithaca: Cornell University Press, 1989), 97, 103–4.

4. Definition of "squall line" from American Meteorological Society, Online Glossary of Meteorology, at https://glossary.ametsoc.org/wiki/Squall_line/.

5. Friedman, *Appropriating the Weather*, 93–94.

6. "Weather and War," *Straits Times*, 18 December 1916, 12. See also UK Meteorological Office, "The Met Office in the Field" at https://www.metoffice.gov.uk/binaries/content/assets/metofficegovuk/pdf/research/library-and-archive/library/publications/factsheets/remember_world-war-one-and-two.pdf/.

7. The only major conflict in Malaya was a naval battle that took place off the coast of Penang Island in 1914. J. R. Robertson, *The Battle of Penang: World War One in the Far East* (Kuala Lumpur: Trafalgar Publishing House, 2012).

8. "Selangor administration Report for 1891," 607/1892, NAM.

9. Malcolm Walker, *History of the Meteorological Office* (Cambridge: Cambridge University Press, 2011), 203–4.

10. S. C. M. Paine, *The Wars for Asia, 1911–1949* (Cambridge: Cambridge University Press, 2012); Christopher M. Bell, "'Our Most Exposed Outpost': Hong Kong and British Far Eastern Strategy, 1921–1941," *Journal of Military History* 60:1 (1996): 61–88; William Roger Louis, *British Strategy in the Far East, 1919–1939* (Oxford: Clarendon Press, 1971).

11. Gillian R. Topley, *British Naval Policy in the 1920s* (masters' thesis, University of Adelaide, 2000), ch. 3; Malcolm H. Murfett, John N. Miksic, Brian P. Farrell, and Ming Shun Chiang, *Between Two Oceans: A Military History of Singapore from 1275 to 1971*, 2nd ed. (Singapore: Marshall Cavendish Editions, 2011), ch. 6.

12. Viscount Wimborne and Marquess of Salisbury, "New Naval Base at Singapore, debated 5 June 1923," Hansard: Lords Chamber, vol. 54, columns 399–403.

13. Martin Mahony, "For an Empire of 'All Types of Climate': Meteorology as an Imperial Science," *Journal of Historical Geography* 51 (2016): 34. For an interesting take on futurism and aviation in popular culture, see Michael McCluskey and Luke Seaber, eds., *Aviation in the Literature and Culture of Interwar Britain* (Cham, Switzerland: Palgrave MacMillan, 2021). See also Henry, "Assembling the Weather"; Cushman, "Struggle over Airways."

14. Paul N. Edwards, *A Vast Machine: Computer Models, Climate Data, and the Politics of Global Warming* (Cambridge, MA: MIT Press, 2010), 54.

15. Martin Mahony, "Historical Geographies of the Future: Airships and the Making of Imperial Atmospheres," *Annals of the American Association of Geographers* 109:4 (2019): 1284–85.

16. "Air Travel in the Future," *Singapore Free Press and Mercantile Advertiser*, 5 December 1928, 12.

17. "The 'Flivver' of the Air," *Singapore Free Press and Mercantile Advertiser*, 23 September 1926, 8.

18. "Air Liners of the Future," *Singapore Free Press and Mercantile Advertiser*, 19 October 1927, 6; "Ceylon–England in Seven Days," *Singapore Free Press and Mercantile Advertiser*, 16 November 1927, 15; "First Real Liner of the Air," *Malaya Tribune*, 19 November 1927, 4; "Wonders of R100," *Malaya Tribune*, 12 January 1928, 3.

19. Mahony, "Historical Geographies," 1295.

20. Anon., "Arrival at Singapore," *Pinang Gazette and Straits Chronicle*, 5 December 1919, 2. "Ross M. Smith standing by his aircraft at Singapore en route for Australia," Classmark Y30311B, Lillian Newton Photograph Collection on Singapore, RCS, CUL.

21. Anon., "The Red Rose Arrives," *Straits Times*, 7 January 1928, 3; Anon., "The 'Red Rose' Starting from London Arrives in Singapore," *Malayan Saturday Post*, 15 January 1928, 13; Anon., "A wilted 'Red Rose,'" *Singapore Free Press and Mercantile Advertiser*, 17 January 1928, 9. See also Chrystopher J. Spicer, *The Flying Adventures of Jessie Keith "Chubbie" Miller: The Southern Hemisphere's First International Aviatrix* (Jefferson, NC: McFarland, 2017), 6, 61–62.

22. "Miss Amy Johnson Here To-day," *Singapore Free Press and Mercantile Advertiser*, 17 May 1930, 10. For Hollywood: Robert Wohl, "Republic of the Air," *Wilson Quarterly* 17:2 (1993): 106–18. See also Gabriele D'Annunzio, *Yes, Perhaps No* (1910); C. D. Burney, *The World, the Air, and the Future* (1929); Antoine Saint-Exupery, *Night Flight* (1931). For RKO: "Amusements," *Malaya Tribune*, 6 December 1934, 6; "This Week's Films," *Straits Times*, 15 October 1939, 5.

23. "The Java-Malaya Flight," *Singapore Free Press and Mercantile Advertiser*, 12 February 1930, 3.

24. Lucille Yap, "Singapore Takes Flight: A Hundred Years of Aviation," *Be-Muse* 4:2 (2011), at https://www.roots.gov.sg/stories-landing/stories/singapore-takes-flight-a-hundred-years-of-aviation/story/.

25. P. Hutton, *Wings over Singapore: The Story of Singapore Changi Airport* (Singapore: MPH Magazines, 1981), 23.

26. *Straits Settlements Government Gazette*, 17 April 1930, 660.

27. *Report of the Tenth Meeting of the International Meteorological Committee*, Authority of the Meteorological Committee, 1913, 71.

28. Anon., "By The Way," *Straits Times*, 11 January 1902, 5. One wonders, however, to what extent the "friend's" observations were themselves accurate. It should also be noted that the hospital was situated in the eastern downtown core, whereas Bukit Timah was still relatively rural at this time, as well as being slightly elevated, facts that may account for the greater daily temperature fluctuation because of better air flow.

29. Anon., "Feet of Rain," *Singapore Free Press and Mercantile Advertiser (Weekly)*, 3 January 1907, 4; "The Agricultural Bulletin," *Straits Times*, 28 September 1911, 6.

30. "Rainfall," *Malaya Tribune*, 16 May 1917, 7.

31. The Gassiot Committee of the Royal Society administered funds donated for meteorological and magnetical observations by John Gassiot in 1871. See D. C. Martin, "The Gassiot Committee of the Royal Society and Meteorological Research," *Nature* 190:4772 (1961): 212–13; E. Gold, *William Napier Shaw, 1854–1945: Obituary* (Royal Society Publications: 1945), 203–30.

32. Edwards, *Vast Machine*, 57–58; Ernest Gold, "Obituary: William Napier Shaw, 1854–1945," *Biographical Memoirs of Fellows of the Royal Society* 5:14 (1945): 212–13.

33. Aitor Anduaga, "Transnational Co-production of Knowledge: The Standardisation of Typhoon Warning Codes in the Far East, 1900–1939," *Minerva* (2022): 7–9.

34. "Local and Personal," *Straits Times*, 9 April 1917, 8.

35. Napier Shaw to E. A. Grindle at the Colonial Office, "Conference of Meteorologists of the Empire to be held in London in May 1919: proposal to include dominion and colonial meteorological delegates at the conference," ff. 397–401, CO323/812/54/187, TNA.

36. Sir George Maxwell and Herbert C. Robinson, A Meteorological Department for Malaya, 14 August 1927, 1, CO273/541/4, TNA.

37. "Naval Signalmen for Malaya," *Malaya Tribune*, 28 September 1920, 7; "That Hill Station," *Malaya Tribune*, 25 June 1920, 4.

38. "Observers for Gunong Tahan Meteorological Station, 1921," SEL: SEC 1321/1921, 1921, NAM; "Preliminary Report on the Meteorological Observations: Taken on Gunong Tahan: July 1921–June 1923," SEL: SEC 4451/1923, 1923, NAM.

39. Geoffrey K. Pakiam, "'Not the Oil of the Country': Smallholders and British Malaya's Oil Palm Industry, 1929–1941," *Agricultural History* 95:1 (2021): 70.

40. Joseph Hodge, *Triumph of the Expert: Agrarian Doctrines of Development and the Legacies of British Colonialism* (Athens: Ohio University Press, 2007), 11; Martin Mahony, "Weather, Climate, and the Colonial Imagination: Meteorology and the End of Empire," in Martin Mahony and Samuel Randalls, eds., *Weather, Climate, and the Geographical Imagination: Placing Atmospheric Knowledges* (Pittsburgh, PA: University of Pittsburgh Press, 2020), 171.

41. Timothy P. Barnard and Joanna W. C. Lee, "A Spiteful Campaign: Agriculture, Forests, and Administering the Environment in Imperial Singapore," *Environmental History* 27:3 (2022): 467–90; Jeyamalar Kathirithamby-Wells, "The Implications of Plantation Agriculture for Biodiversity in Peninsula Malaysia: A Historical Analysis," in *Beyond the Sacred Forest: Complicating Conservation in Southeast Asia*, ed. Michael R. Dove, Percy E. Sajise, and Amity A. Dolittle (Durham: Duke University Press, 2011), 63.

42. Mahony, "All Types of Climate," 36; BAAS, *Report of the Seventy-Sixth Meeting held in South Africa, 1905*, 86.

43. Quoted from Friedman, *Appropriating the Weather*, 109.

44. Kathirithamby-Wells, "Implications of Plantation Agriculture," 71.

45. J. H. Drabble, "The Plantation Rubber Industry in Malaya up to 1922," *JMBRAS* 40:1 (1967): 63; Kathirithamby-Wells, "Implications of Plantation Agriculture," 66; Lynn Hollen Lees, *Planting Empire, Cultivating Subjects: British Malaya, 1786–1941* (Cambridge: Cambridge University Press, 2017), 176; A. W.

King, "Plantation and Agriculture in Malaya, with Notes on the Trade of Singapore," *Geographical Journal* 93:2 (1939): 136, 141.

46. *First Annual Report of the Colonial Research Committee, for the period ended 31st December 1920* (London: HMSO, 1921).

47. *First Annual Report of the Colonial Research Committee*, 17 (appendix).

48. Timothy P. Barnard, *Nature's Colony: Empire, Nation and Environment in the Singapore Botanic Gardens* (Singapore: NUS Press, 2016), 14.

49. "Colonial Office Parley on Soil Erosion," *China Mail*, 26 July 1938.

50. "Soil Erosion (Part 1). Commission of Research: Africa, 1st Session, 2 January 1942," f. 3, *Remarks submitted by the Members of the Commission and others on the Proceedings of the First Three Sessions*, 9, CO852/394/14, TNA.

51. *Remarks submitted by the Members of the Commission and others on the Proceedings of the First Three Sessions*, 18, 20.

52. Michael Worboys, "British Colonial Science Policy, 1918–1939," in *Colonial Sciences: Researchers and Institution*, ed. Patrick Petitjean (Paris: Orstom Editions, 1996), 108.

53. "Research Institute," *Malaya Tribune*, 20 April 1928, 7. See also "Soil Erosion in Malaya," *Straits Times*, 20 June 1928, 9; *SSGG: Annual Report for the Survey Department, Straits Settlements, for the Year 1929* (Singapore: Government Printing Office, 1938), 1. The headquarters were moved to Singapore in 1931. *SSGG: Annual Report on the Survey Department, Straits Settlements for the Year 1932*, 2.

54. "Compilation and Issue of Meteorological Returns for the FMS by the Museums Dept.," SEL: SEC 1108/1925, 1925, NAM.

55. TNA, CO273/541/4, "Meteorology in Malaya," Letter from the High Commissioner F.M.S. to the Colonial Office, 14 October 1927, pp. 1-5, 2; *SSGG: Annual Report on the Survey Department, Straits Settlements for the Year 1932*, 1–3, 2; *SSGG: Annual Report for the Survey Department, Straits Settlements, for the Year 1929*, 1, CO273/541/4, TNA. The headquarters were moved to Singapore in 1931. *SSGG: Annual Report on the Survey Department, Straits Settlements for the Year 1932*, 2.

56. Robert L. Jarman, ed., *Annual Reports of the Straits Settlements 1855–1941*, 12 vols. (Slough: Archive Editions, 1998), 9:304. See also "Meteorology in Malaya," file 10, 1927, ff. 1–24, CO273/541/4, TNA.

57. "Meteorology in Malaya," file 10, 1927, ff. 1r, 3r, CO273/541/4, TNA.

58. Walter Makepeace, Gilbert E. Brooke, Roland St. John Braddell, eds., *One Hundred Years of Singapore* (London: John Murray, 1921), 1:481; "Equipment of Observatory at Mount Faber," RGO 8/93 1928, 1–5, CUL.

59. "Supplement to the Straits Settlements' Government Gazette, 31 August 1928," *SSGG* (Singapore: Government Printing Office, 1928), 3:8; "Apparatus for Installation at Mount Faber Observatory, 1928," CO273/552/12, TNA; "Equipment of Observatory at Mt. Faber, 1928," ff. 1–5, RGO 8/93 1928, CUL.

60. Anon., "Wireless Storm Warnings," *Straits Times*, 29 May 1903, 6.

61. British Pathé: "Opening of Imperial Wireless Station at Leafield, 1921," at https://www.britishpathe.com/asset/168972/.

62. Jill Hills, *The Struggle for Control of Global Communication: The Formative Century* (Urbana: University of Illinois Press, 2010), 224–29; Hansard 1803–2005, Commons Sitting, Post Office, 13 July 1922.

63. Hugh Clifford, Governor of the Straits Settlements to the Rt. Hn. L. C. M. S. Amery MP, Colonial Office, 25 July 1928, Mount Faber Instruments, RGO 8/93, CUL.

64. Jeon, "Flying Weather Men," 119.

65. For more on the East Asian observatory networks, see Robert Bickers, "'Throwing Light on Natural Laws': Meteorology on the China Coast, 1869–1912," in R. Bickers and I. Jackson, eds., *Treaty Ports in Modern China: Law, Land and Power* (London: Routledge, 2016), 179–200; Marlon Zhu, "Typhoons, Meteorological Intelligence, and the Inter-Port Mercantile Community in Nineteenth-Century China" (unpublished PhD thesis, Department of History, Binghamton University, State University of New York, 2012); Kevin MacKeown, *Early China Coast Meteorology: The Role of Hong Kong* (Hong Kong: Hong Kong University Press, 2011); P. Y. Ho, *Weathering the Storm: Hong Kong Observatory and Social Development* (Hong Kong: Hong Kong University Press, 2003).

66. David Lambert and Alan Lester, eds., *Colonial Lives across the British Empire: Imperial Careering in the Long Nineteenth Century* (Cambridge: Cambridge University Press, 2006), 8.

67. Malayan Meteorological Service, *Annual Report 1948*, 1–2, 4, Federal Secretariat (FS) 13173/1949, NAM; *SSGG: Annual Report on the Survey Department, Straits Settlements for the Year 1932*, 2; *Annual Report of the Malayan Meteorological Service, 1949* (Singapore: Government Printing Office, 1950), appendix.

68. Untitled article, *Malaya Tribune*, 10 December 1936, 20

69. Ivory Kedit, "Communities of Singapore (Part 3)," accession no. 001249, reel 1, pp. 1–11, Oral Histories, NAS.

70. "Social and Personal," *Pinang Gazette and Straits Chronicle*, 10 November 1936, 4. See also Untitled article, *Malaya Tribune*, 13 February 1937, 10; Untitled article, *Malaya Tribune*, 30 August 1938, 6.

71. Kanagasabai s/o Narayasamy (N. K. Sabai), "The Public Service," accession no. 001496, reel 1, pp. 3–5, Oral Histories, NAS.

72. *SSGG: Annual Report for the Survey Department, Straits Settlements, for the Year 1929*, 1; "Conference of Empire Meteorologists," *Nature* 3124:124 (1929): 427–28.

73. Correspondence relating to the Conferences of Directors 1929–30, Notes by M. Gent, 1 August 1930, ff. 3–4, CO825/8/5, TNA.

74. International Meteorological Committee and Far East Regional Committee. Letter to W. G. A Ormsby Gore from Government House, Singapore, No. 299, 21 August 1937, CO323/1525/1, TNA.

75. International Meteorological Committee and Far East Regional Committee, *Preliminary Report of the First Conference in Hong Kong [of the Regional Committee of the International Meteorological Organisation], January 13–21, 1937*, 6–8, CO323/1525/1, TNA.

76. International Meteorological Committee and Far East Regional Committee, *Preliminary Report of the First Conference in Hong Kong*, 24.

77. International Meteorological Committee and Far East Regional Committee, *Preliminary Report of the First Conference in Hong Kong*, 29.

78. International Meteorological Committee and Far East Regional Committee, *Preliminary Report of the First Conference in Hong Kong*, 38.

79. Anduaga, "Transnational Co-production," 12.

80. Anduaga, "Transnational Co-production," 53.

81. Edwards, *Vast Machine*, 56.

82. Paul H. Kratoska, *The Japanese Occupation of Malaya and Singapore, 1941–45: A Social and Economic History* (Singapore: NUS Press, 2018), 21.

83. Christopher Bayley and Tim Harper, *Forgotten Armies: The Fall of British Asia, 1941–1945* (Cambridge, MA: Harvard University Press, 2005), 33.

84. Margaret Shennan, *Out in the Midday Sun: The British in Malaya, 1880–1960* (Singapore: Monsoon Books, 2015), 294, 300.

85. H. B. F. Moorhead, "Storm and Hurricane Warnings," *Royal Gazette and Colonist Daily*, 23 August 1935, 9; Anon., "Malaya's New Weather Chief Was in Navy," *Straits Budget*, 15 December 1938, 13.

86. *Annual Report of the Malayan Meteorological Service*, 1949 (Singapore: Government Printing Office, 1950), appendix.

87. Anon., "Weather Forecasts Vital in Modern Warfare," *Malaya Tribune*, 18 June 1940, 5.

88. Wen-Qing Ngoei, *Arc of Containment: Britain, the United States, and Anticommunism in Southeast Asia* (Ithaca: Cornell University Press, 2019), 19–20.

89. Kratoska, *Japanese Occupation*, 30.

90. Kratoska, *Japanese Occupation*, 31, 34.

91. Regardless, prior reconnaissance enabled Japanese meteorologists to gain a great many observations and information on Malayan weather: UEDA Kunitaro, "Japanese Occupation of Singapore," accession no. 001444, Oral Histories, NAS; *Annual Report of the Malayan Meteorological Service, 1949* (Singapore: Government Printing Office, 1950), 3.

92. UEDA Kunitaro, "Japanese Occupation of Singapore," accession no. 001444, Oral Histories, NAS.

93. The continued support for science in Malaya is the key theme of the autobiographical account of the Japanese occupation of Singapore by E. J. H. Corner, assistant director of the Singapore Botanic Gardens. See E. J. H. Corner, *The Marquis: A Tale of Syonan-to* (Singapore: Heinemann Asia, 1981). See also Barnard, *Nature's Colony*, 211–12.

94. Kanagasabai s/o Narayasamy (N. K. Sabai), "The Public Service," accession no. 001496, reel 1, pp. 3–5, Oral Histories, NAS.

95. The author wishes to thank Michael Pether for sharing his information on Moorhead's internment at the Palembang and Muntok internment camps during the Second World War.

PART II: THINKING WITH WEATHER

1. John Cameron, *Our Tropical Possessions in Malayan India* (London: Smith Elder and Co., 1865), 154–55.

2. Nigel P. Taylor, "The Environmental Significance of the Singapore Botanic Gardens," in *Nature's Colony: Empire, Nation and Environment in the Singapore Botanic Gardens*, ed. Timothy P. Barnard (Singapore: NUS Press, 2016), 116; Barnard, *Nature's Colony*, 14–49.

3. Barnard, *Nature's Colony*, 114.

4. Lewis Pyenson, *Empire of Reason: Exact Sciences in Indonesia* (Leiden: Brill, 1989). For critique, see Paolo Palladino and Michael Worboys, "Science and Imperialism," *Isis* 84 (1993): 91–102.

CHAPTER 3: THE GREAT DESICCATION DEBATE

1. C. J. Glacken, *Traces on the Rhodian Shore* (Berkeley: University of California, 1967), 129–30.

2. J. Needham, *Science and Civilisation in China* (Cambridge: Cambridge University Press, 1980), vol. 5, pt. 28, 245, quoted in Richard Grove, *Green Imperialism: Colonialism, Tropical Island Edens and the Origins of Environmentalism, 1600–1860* (Cambridge: Cambridge University Press, 1995), 62.

3. Stephen Legg, "Debating the Climatological Role of Forests in Australia, 1827–1949: A Survey of the Popular Press," in *Climate, Science, and Colonization: Histories from Australia and New Zealand*, ed. James Beattie, Emily O'Gorman, and Matthew Henry (New York: Palgrave Macmillan, 2014), 119.

4. Richard Grove, *Ecology, Climate and Empire: Colonialism and Global Environmental History, 1400–1940* (Cambridge, UK: White Horse Press, 1997), 5.

5. Joachim Radkau, *Nature and Power: A Global History of the Environment*, trans. Thomas Dunlap (Washington, DC: German Historical Institute, 2008), 164–68.

6. Brett M. Bennett, *Plantations and Protected Areas: A Global History of Forest Management* (Cambridge, MA: MIT Press, 2015), 36.

7. Alexander von Humboldt, *Cosmos: A General Survey of Physical Phenomena of the Universe* (London: Hippolyte Baillière, 1845), vol. 1.

8. Fabien Locher and Jean-Baptiste Fressoz, "Modernity's Frail Climate: A Climate History of Environmental Reflexivity," *Critical Inquiry* 38 (2012): 585.

9. Bennett, *Plantations and Protected Areas*, 37.

10. A. C. Becquerel, "Memoir upon Forests, and Their Climatic Influence," reprinted in Franklin Benjamin Hough, *Report on Forestry* (Washington, DC: Government Printing Service, 1878), 310–32. For example, "The Value of Trees," *Pall Mall Gazette* 103 (13 June 1865): 10; *Prairie Farmer* 18 (13 October 1866): 242, both in Part I: 1800–1900, British Library Newspapers (BLN).

11. Becquerel, "Memoir upon Forests," 328–30.

12. BAAS, "Report of a Committee Appointed by the British Association to Consider the Probable Effects in an Economic and Physical Point of View of the Destruction of Tropical Forests, 1852," *Twenty-First Meeting of the British Association for the Advancement of Science 1851* (London: John Murray, 1852), 79.

13. BAAS, "Report of a Committee," 101.

14. Bennett, *Plantations and Protected Areas*, 53–54.

15. Stephen Legg suggests that Marsh's comments on desiccation were often taken out of context or exaggerated by desiccation advocates. Legg, "Debating the Climatological Role of Forests," 123.

16. George Perkins Marsh, *The Earth as Modified by Human Action: A New Edition of Man and Nature* (New York, 1877), 152.

17. For discussion of Marsh and his work, see Lowenthal's introduction in David Lowenthal, ed., *George Perkins Marsh, Man and Nature: Or, Physical Geography as Modified by Human Action, 1864* (Seattle: University of Washington Press, 2003).

18. George Perkins Marsh, "The Study of Nature" (1860), reproduced in Stephen C. Trombulak, ed., *So Great a Vision: The Conservation Writings of George Perkins Marsh* (Hanover: Middlebury College Press, University Press of New England, 2001), 86.

19. Marsh, *Earth as Modified by Human Action*, vi.

20. G. L. Andrew, *On the Sanitary Value of Forests* (Cambridge, MA, 1879), 1.

21. "Forest Preservation in the Colonies," *Pall Mall Gazette* 4320 (26 December 1878): 2183, Part I: 1800–1900, BLN.

22. John Croumbie Brown, *Forests and Moisture; or the Effects of forests on humidity of climate* (Edinburgh: Oliver and Boyd, 1877).

23. Morag Bell, *Geography and Imperialism, 1820–1940* (Manchester: Manchester University Press, 1995), 47.

24. Brown, *Forests and Moisture*, 307.

25. Deborah Coen, *Climate in Motion: Science, Empire, and the Problem of Scale* (Chicago: Chicago University Press, 2018), 248–51.

26. Marsh, *Earth as Modified by Human Action*, 155–56.

27. Donald J. Pisani, "Forests and Conservation, 1865–1890," *Journal of American History* 72:2 (1985): 350.

28. B. E. Fernow, "Notes on the Sanitary Significance of Forests," *US Department of Agriculture, Forestry Division, Bulletin* No. 7 (Washington, 1893): 171–86;

E. Ebermeyer, *The physical effects of the forest on air and soil and their Climatological and Hygienic Importance* (1873).

29. James M. Anders, "Forests—Their Influence upon Climate and Rainfall," *American Naturalist* 16:1 (1882): 19–30.

30. James M. Anders, *Sanitary Influences of Forest Growth* (Philadelphia, 1885).

31. For example, Anders's work was reprinted and discussed in A. M. and J. Ferguson, eds., "Influence of Forests on Rainfall," *Tropical Agriculturalist and Magazine of the Ceylon Agricultural Society* (1882): 952.

32. H. R. Hopkins, "Forestry and the Public Health," *American Forestry* 22 (1916): 417–18.

33. Henry M. Fisher, "The Evergreen Forest Air Treatment of Pulmonary and other Diseases," *Forest Leaves* 1 (1886): 6.

34. Fisher, "Evergreen Forest Air Treatment," 8.

35. For Francis Reid, see Edmund Alexander Parkes, *A Manual of Practical Hygiene: Intended Especially for Medical Officers of the Army and for Civil Medical Officers of Health* (J&A Churchill, 1873), 616–17.

36. Andrew, "On the Sanitary Value of Forests," 3.

37. "Forest Preservation in the Colonies," *Pall Mall Gazette* 4320, 26 December 1878, 2183, and "Forest Preservation in the Colonies," *Freeman's Journal and Daily Commercial Advertiser* (Dublin, Ireland), Saturday, December 28, 1878, n.p., both in Part I: 1800–1900, BLN.

38. Glacken, *Traces on the Rhodian Shore*, 485, 490.

39. F. B. Hough, "Public Health Interests concerned in the Preservation of Certain Primeval Forest and in the Cultivation of Groves and Trees," *American Public Health Association Report, 1875* (1875): 177.

40. Hugh Williamson, "An Attempt to Account for the Change of Climate, which has been Observed in the Middle Colonies of North America," *Transactions of the American Philosophical Society*, 1 (1769–1771): 277, 280.

41. A good overview of the changing views in a British urban context can be found in Stephen Halliday, "Death and Miasma in Victorian London: An Obstinate Belief," *British Medical Journal* 323:7327 (2001): 1469–71.

42. Legg, "Debating the Climatological Role of Forests," 126–28.

43. Prince Kropotkin, "The Desiccation of Eur-Asia," *Geographical Journal* 23:6 (1904): 722–34.

44. Kropotkin, "The Desiccation of Eur-Asia," 725.

45. "Forest Conservancy in India," *Morning Post* (London, England), issue 33880, Tuesday, January 25, 1881, 6, Part II: 1800–1900, BLN.

46. John Augustus Voelcker, *Report of the Improvement of Indian Agriculture* (London: Eyre and Spottiswoode, 1893), 30.

47. Jeyamalar Kathirithamby-Wells, *Nature and Nation: Forests and Development in Peninsula Malaysia* (Singapore: Nordic Institute of Asian Studies and

NUS Press, 2005), 74; Gregory A. Barton, "Empire Forestry and the Origins of Environmentalism," *Journal of Historical Geography*, 27:4 (2001): 530–33.

48. For more on this subject in the Singapore context, see Tony O'Dempsey, "Singapore's Changing Landscape since c. 1800," in *Nature Contained: Environmental Histories of Singapore*, ed. Timothy P. Barnard (Singapore: NUS Press, 2014), 17–48; Richard T. Corlett, "The Ecological Transformation of Singapore, 1819–1990," *Journal of Biogeography* 19:4 (1992): 411–20.

49. T. M. Ward, "Contributions to the Medical Topography of Prince of Wales Island, Or Pulo Pinang," in T. M. Ward and J. P. Grant, *Official Papers on the Medical Statistics and Topography of Malacca and Prince of Wales Island and on the Prevailing Diseases of the Tenasserim Coast* (Pinang: Government Press, 1830), 61. Sourced from the Wellcome Collection online at https://wellcomecollection.org/works/uhqkbwe2.

50. Ward, "Contributions to the Medical Topography," 61.

51. Walter Makepeace, Gilbert Brooke, and Roland St. John Braddell, *One Hundred Years of Singapore*, 2 vols. (London: John Murray, 1921), 2:308; Anon., "The Timber Trees of the Straits Settlements," *Straits Times Overland Journal* 18 (April 1881): 2.

52. H. Price, "Planting in Singapore" in Makepeace, Brooke, and Braddell, *One Hundred Years of Singapore*, 2:83.

53. Grove, *Green Imperialism*, 437–40.

54. "Biographical memoir of Captain. T. J. Newbold," *Journal of Straits Branch of the Royal Asiatic Society* 19 (1887): 143–52.

55. T. J. Newbold, "On the temperature of the springs, wells and rivers of India and Egypt, and of the sea and table lands within the tropics; with a few remarks on M. Boussingault's mode of ascertaining the mean temperature of equinoctial regions," *Proceedings of the Royal Society of London* 5 (1851): 502–3.

56. T. J. Newbold, "Notice of River Dunes on the Banks of the Hogri and Pennaur," *Madras Journal of Literature and Science* 9 (1839): 309–10.

57. Despite Oxley's medical qualifications, he is best known for establishing one of the first nutmeg plantations in Singapore, which became the Oxley Estate. Gretchen Liu, *Singapore: A Pictorial History, 1819–2000* (Richmond: Curzon Press, 2001), 20.

58. J. T. Thomson, "General Report on the Residency of Singapore, drawn up principally with a view of illustrating its agricultural statistics," *Journal of the Indian Archipelago and Eastern Asia* 3 (1849): 618–25.

59. James Low, *A Dissertation on the Soil & Agriculture of the British Settlement of Penang, Or Prince of Wales Island, in the Straits of Malacca: Including Province Wellesley on the Malayan Peninsula. With Brief References to the Settlements of Singapore & Malacca* (Singapore Free Press Office, 1836), 817–18. Low's work is referred to and commented upon by "A Correspondent" in "Capt. Low's Dissertation," *Singapore Free Press and Mercantile Advertiser*, 20 October 1836, 2.

60. J. R. Logan, "The Probable Effects on the Climate of Pinang of the Continued Destruction of its Hill Jungles," *Journal of the Indian Archipelago and Eastern Asia* 2 (1848): 534.

61. "Maingay, Alexander Carroll (1836–1869)," Global Plants database, hosted by JSTOR, at https://plants.jstor.org/stable/10.5555/al.ap.person.bm000327081/.

62. A. C. Maingay, "Report on the Timber and Forest Conservancy of Malacca, 20 August 1865," *SSGG*, 1 September 1865, 388.

63. "Our Forest Conservancy," *Straits Times*, 9 September 1865, 1.

64. Anon., untitled article, *Singapore Free Press and Mercantile Advertiser*, 14 September 1865, 2.

65. "Editorial," *Singapore Free Press and Mercantile Advertiser*, 28 September 1865, 2.

66. "Editorial," *Singapore Free Press and Mercantile Advertiser*, 28 September 1865, 2.

67. "Editorial," *Singapore Free Press and Mercantile Advertiser*, 28 September 1865, 2.

68. Historian C. M. Turnbull notes that, although the *Straits Times* and the *Singapore Free Press* were two of the more important papers of the period, they were both underfunded and understaffed. The editors were forced to manage the papers essentially as a hobby, while they supported themselves financially from other business interests. C. M. Turnbull, "The European Mercantile Community in Singapore, 1819–1867," *Journal of Southeast Asian History* 10:1 (1969): 19.

69. Fiona Williamson and Clive Wilkinson, "Asian Extremes: Experience and Exchange in the Development of Meteorological Knowledge, c. 1840–1930," *History of Meteorology* 8 (2017): 159–78; Ricardo García-Herrera, C. Wilkinson, F. B. Koek, M. R. Prieto, N. Calvo, and E. Hernández, "Description and General Background to Ships' Logbooks as a Source of Climatic Data," *Climatic Change* 73:1 (2005): 13–36.

70. John Cameron, *Our Tropical Possession in Malayan India* (London, 1865), 152.

71. The paper had been under the editorship of Abraham Logan, J. R. Logan's brother, until Abraham's retirement in January 1865. See "Mr Logan," *Straits Times*, 7 January 1865, 1.

72. Jonas Daniel Vaughan, at http://eresources.nlb.gov.sg/infopedia/articles/SIP_1826_2011-08-11.html/.

73. Vaughan's observations were made at his estate on River Valley Road and published in the *Straits Settlements Government Gazette* from 1862 to 1869. See, for example, *Singapore Government Gazette* (1865): 20, 278, 419, C. J. Koh Law Library, 276/1, NUS. For more information on these and other early observations for the country, see Elaine Gao, Bertrand Timbal, and Fiona Williamson, "Creat-

ing Singapore's Longest Monthly Rainfall Record from 1839 to the Present," *MSS Research Letters* 2 (2018): 3–11.

74. J. J. L. Wheatley, "Notes on the Rainfall of Singapore," *Journal of the Straits Branch of the Royal Asiatic Society* (Singapore, 1881): 31–50. Wheatley is also noted in Singapore in 1871, as superintendent apothecary for the lunatic asylum. Beng Yeong Ng, *Till the Break of Day: A History of Mental Health Services in Singapore, 1841–1993* (Singapore: NUS Press, 2016), 14; *Chronicle and Directory for China, Corea, Japan etc., for the year 1888* (Hong Kong, 1888), 270. In later years, Wheatley became surgeon on the SS *Hong Moh*. He died at sea in 1909. Makepeace, Brooke, and Braddell, *One Hundred Years of Singapore*, 2:273.

75. A. M. Skinner, "Straits Meteorology," *JMBRAS* 12 (1883): 245–55.

76. A. C. Maingay, "Meteorology of the Straits," *Straits Times*, 18 December 1865.

77. "Maingay, Alexander Carroll (1836–1869)," Global Plants database, hosted by JSTOR, at https://plants.jstor.org/stable/10.5555/al.ap.person.bm000327081/.

78. See, for example, a series of anonymous articles addressed to the editor: "Effects of Forest Denudation," *Straits Times*, 10 January 1874, 5; "Deforestation," *Straits Times*, 25 May 1878, 3; "The Rainfall," *Straits Times*, 29 June 1878, 5; "Forest Tree Planting and Its Influence on Climate," *Straits Times*, 20 July 1878, 1.

79. H. L. Randall, "Annual Medical Report for the Straits Settlements, 1873," in *SSGG*, 2 May 1874, 247.

80. D. Singh, R. Seager, B. Cook, M. Cane, M. Ting, E. Cook, and M. Davis, "Climate and the Global Famine of 1876–78," *Journal of Climate* 31 (2018): 9445–67.

81. J. F. A. McNair, "Report by the Colonial Engineer on the timber forests in the Malayan Peninsula, 21 June 1879," *SSGG*, 3 October 1879, 903, CO276/10, TNA.

82. Report of H. J. Murton, 21 December 1877, presented before the Legislation Council, 15 March 1878, Sessional Papers, ff. 21–25v, CO275/22, TNA.

83. Report of H. J. Murton, 21 December 1877, 22v.

84. "Copy of Colonial Engineers Minute on Survey," dated 27 August 1878, *SSGG*, 3 October 1879, 893, CO276/10, TNA.

85. "Copy of Colonial Engineers Minute on Survey," 893.

86. Nathaniel Cantley, "Report on the Forests of the Straits Settlements, 1882," Annual Report on the Botanic Gardens, Singapore, 1884 (Singapore, 1885), 1, at https://www.biodiversitylibrary.org/item/234420#page/75/mode/1up/.

87. Cantley, "Report on the Forests, 1882," 4. For more on the integration of forestry under the Singapore Botanic Garden's purview, see Barnard, *Nature's Colony*, 50–83.

88. Cantley, "Report on the Forests, 1882," 5. Cantley's annex contains a map of forest reserves, which incidentally highlights the Thomson Road area. The

road is not named but is quite clear, running north across the eastern edge of the Waterworks' Reservoir. N. Cantley, *Map of the Island of Singapore. Annex to Report on the Forests of the Straits Settlements,* 1882, at http://search.nlb.gov.sg/spatialdiscovery/.

89. His evidence is borne out by the rainfall statistics for 1870–1885 presented in a later annual meteorological report. T. Irvine Rowell, "Chart showing the Range of Mean Annual Rainfall at Singapore from 1870 to 1885," and "Chart showing the Range of Mean Annual number of rainy days at Singapore from 1870 to 1885," *Annual Report of the Meteorological Observations of the Straits Settlements for the year 1884–1885* (Singapore, 1886), appendix.

90. "Map of the Island of Singapore," Annexure to Report on the Forests of the Straits Settlements, 1882, by Nathaniel Cantley, media image no. 20050000974-0093_TM000020_000028_TM. NAS.

91. T. Irvine Rowell, "Meteorological Report, for the year 1885," *Journal of the Straits Branch of the Royal Asiatic Society* 16 (1885): 390.

92. *SSGG*, 12 May 1885, 723.

93. Irvine Rowell, "Meteorological Report, for the year 1885," 390.

94. J. J. L. Wheatley, "Notes on the Rainfall of Singapore," *Journal of the Straits Branch of the Royal Asiatic Society* (Singapore, 1881): 31–50.

95. Wheatley, "Notes on the Rainfall of Singapore," 31.

96. Wheatley, "Notes on the Rainfall of Singapore," 38–39.

97. Katherine Anderson, "Looking at the Sky: The Visual Context of Victorian Meteorology," *British Journal for the History of Science* 36:3 (2003): 304.

98. Emily Sadka, "The Journal of Sir Hugh Low, Perak, 1877," *JMBRAS* 27:4 (1954): 5–108.

99. Kathirithamby-Wells, *Nature and Nation*, 61.

100. Kathirithamby-Wells, *Nature and Nation*, 61.

101. Legg, "Debating the Climatological Role of Forests," 124.

102. Kathy Willis and Carolyn Fry, *Plants: From Roots to Riches* (London: Hachette UK, 2014), n.p.

103. D. Brandis, Office of the Inspector General of Forests, Calcutta, to Sir Joseph Dalton Hooker, 15 January 1878, Director's Correspondence 153, ff. 127–31, Royal Botanic Gardens, Kew.

104. For more on what was a struggle of politics and personality, see Timothy P. Barnard and Joanna W. C. Lee, "A Spiteful Campaign: Agriculture, Forests, and Administering the Environment in Imperial Singapore," *Environmental History* 27:3 (2022): 467–90; Kathirithamby-Wells, *Nature and Nation*, 71–72.

105. H. C. Hill, *Report on the Present System of Forest Conservancy in the Straits Settlements, with Suggestions for Future Management* (Singapore: Government Printing Office, 1900); *SSGG*, Index for the year 1901, 2170.

106. Anon., "Rainfall and Deforestation," *Singapore Free Press and Mercantile Advertiser*, 15 June 1905, 371.

107. Anon., "Rainfall and Deforestation," *Singapore Free Press and Mercantile Advertiser (Weekly)*, 16 November 1905, 3.

108. "The Effect of Trees on Climate," *Eastern Daily Mail and Straits Morning Advertiser*, 26 March 1906, 3; "Forests and Rainfall," *Straits Times*, 21 November 1908, 3.

109. B. H. F Barnard, "Effect of Clearing Forest on Rainfall," *Agricultural Bulletin of the Straits and Federated Malay States*, new ser., 10 (1911): 269–70. See also Kathirithamby-Wells, *Nature and Nation*, 73; "Trees and Rainfall," *Straits Times*, 29 August 1911, 2.

CHAPTER 4: HOT IN THE CITY

1. Roland Braddell, *The Lights of Singapore* (Methuen, 1934), reprinted (Oxford: Oxford University Press, 1982), 18, 20.

2. Rohan Deb Roy, *Malarial Subjects: Empire, Medicine and Nonhumans in British India, 1820–1909* (Cambridge: Cambridge University Press, 2017); Warwick Anderson, *Colonial Pathologies: American Tropical Medicine, Race, and Hygiene in the Philippines* (Durham, NC: Duke University Press, 2006); Mark Harrison, *Climates and Constitutions: Health, Race, Environment and British Imperialism in India, 1600–1850* (New Delhi: Oxford University Press, 1999); David N. Livingstone, "Tropical Climate and Moral Hygiene: The Anatomy of a Victorian Debate," *British Journal for the History of Science* 32:1 (1999): 93–110.

3. David Arnold, *The Tropics, and the Traveling Gaze: India, Landscape and Science 1800–1856* (Seattle: University of Washington Press, 2006); D. Kennedy, "The Perils of the Midday Sun: Climatic Anxieties in the Colonial Tropics," in *Imperialism and the Natural World*, ed. J. M. MacKenzie (Manchester: Manchester University Press, 1990), 118–40.

4. See, for example, James R. Fleming, *First Woman: Joanne Simpson and the Tropical Atmosphere* (Oxford: Oxford University Press, 2020).

5. Philipp Lehmann, *Desert Edens: Colonial Climate Engineering in the Age of Anxiety* (Princeton, NJ: Princeton University Press, 2022), 16.

6. Russell McCormmach, *Speculative Truth: Henry Cavendish, Natural Philosophy, and the Rise of Modern Theoretical Science* (Oxford: Oxford University Press, 2004), 81; Hasok Chang, *Inventing Temperature: Measurement and Scientific Progress* (Oxford: Oxford University Press, 2004), ch. 1.

7. François Arago, *Biographies of Distinguished Scientific Men*, trans. W. H. Smyth, Rev. Baden-Powell, and Robert Grant (Cambridge: Houghton and Co., 1859), ebook published by Project Gutenberg 2005.

8. "Central Heat of the Terrestrial Globe," in Arago, *Biographies of Distinguished Scientific Men*, 274.

9. Lehmann, *Desert Edens*.

10. See also John McAleer, "Stargazers at the World's End: Telescopes, Ob-

servatories, and 'Views' of Empire in the Nineteenth-Century British Empire," *British Journal for the History of Science* 46:3 (2011): 389–413; Alex Soojung-Kim Pang, *Empire and the Sun: Victorian Solar Eclipse Expeditions* (Stanford, CA: Stanford University Press, 2002), 121–43.

11. J. B. Harley and David Woodward, eds., *The History of Cartography* (Chicago: University of Chicago Press, 1987), xvi.

12. For Ellsworth Huntington, see James Fleming, "Civilization, Climate, and Ozone: Ellsworth Huntington's 'Big' Views on Biophysics, Biocosmics, and Biocracy," in *Weather, Climate and the Colonial Imagination: Placing Atmospheric Knowledges*, ed. Martin Mahony and Samuel Randalls (Pittsburgh: University of Pittsburgh Press, 2020): 215–31.

13. Laura Dassow Walls, *The Passage to Cosmos: Alexander von Humboldt and the Shaping of America* (Chicago: University of Chicago Press, 2009), 128; Frederik Nebeker, *Calculating the Weather: Meteorology in the 20th Century* (California: Academic Press, 1995), 17.

14. Michael Strobl, "Alexander von Humboldt's Climatological Writings," *German Life and Letters* 74:3 (2021): 372.

15. Eberhard Knobloch, "Alexander von Humboldt—The Explorer and the Scientist," *Centaurus* 49 (2007): 3–14, esp. 12.

16. Harley and David Woodward, eds., *The History of Cartography*, xvi.

17. Lehmann, *Desert Edens*, 17. See also Hans von Storch and Nico Stehr, "Eduard Brückner's Ideas: Relevant in His Time and Today," in *Eduard Brückner: The Sources and Consequences of Climate Change and Climate Variability in Historical Times*, ed. N. Stehr and H. von Storch (Netherlands: Springer, 2000), 1–24.

18. Brant Vogel, "The Letter from Dublin: Climate Change, Colonialism, and the Royal Society in the Seventeenth Century," *OSIRIS* 26 (2011): 111–28; Lehmann, *Desert Edens*, introduction; Prince Kropotkin, "The Desiccation of Eur-Asia," *Geographical Journal* 23:6 (1904): 772–34.

19. Julius von Hann, "The Earth as a Whole: Its Atmosphere and Hydrosphere H. R. M.," in "Reviewed Works: Handbook of Climatology by Julius Hann &c," *Geographical Journal* 22:1 (1903): 75–76: Paul N. Edwards, *A Vast Machine: Computer Models, Climate Data, and the Politics of Global Warming* (Cambridge, MA: MIT Press, 2010), 62–66.

20. Blanford was influenced by Norman Lockyer's earlier studies on sunspot cycles. See also George Adamson, "Imperial Oscillations: Gilbert Walker and the Construction of the Southern Oscillation," in *Weather, Climate, and the Geographical Imagination*, ed. Mahony and Randalls, 43–66, esp. 50; Ruth Morgan, "Southern Skies: Australian Atmospheric Research and Global Climate Change," *Disaster Prevention and Management* 30:1 (2021): 47–63.

21. Aitor Anduaga, *Politics, Statistics and Weather Forecasting, 1840–1910: Taming the Weather* (London: Routledge, 2020), esp. ch. 5; Paul N. Edwards, "Meteorology as Infrastructural Globalism," *OSIRIS* 21 (2006): 229–50, esp. 230–31.

22. Martin Mahony, "The 'Genie of the Storm': Cyclonic Reasoning and the Spaces of Weather Observation in the Southern Indian Ocean, 1851–1925," *British Journal for the History of Science* 51:4 (2018): 607–33; Mike Chenoweth, J. M. Vaquero, Ricardo Garcia-Herrera, and Dennis Wheeler, "A Pioneer in Tropical Meteorology: William Sharpe's Barbados Weather Journal, April–August 1680," *Bulletin of the American Meteorological Society* (2007): 1957–64.

23. A. Udías, "Jesuits' Contribution to Meteorology," *Bulletin of the American Meteorological Society* 77:10 (1996): 2307–15.

24. P. Peterson, "Extreme Temperatures in Hong Kong," *Royal Observatory, Hong Kong Technical Note (Local)* 22 (1981): 1–41, esp. 31.

25. Paul N. Edwards, *A Vast Machine: Computer Models, Climate Data, and the Politics of Global Warming* (Cambridge. MA: MIT Press, 2010), 62.

26. Fiona Williamson, "Building a Long-Time Series for Weather and Extreme Weather in the Straits Settlements: A Multi-Disciplinary Approach to the Archives of Societies," *Climate of the Past* 17 (2021): 791–803.

27. William Farquhar, "Appendix I Thermometrical Registers," *Transactions of the Royal Asiatic Society of Great Britain and Ireland* 1:2 (1826): 585–86; John Monckton Coombs, "Appendix II Thermometrical Registers," *Transactions of the Royal Asiatic Society of Great Britain and Ireland* 1:2 (1826), 586–98; "Instructions from the government at Bengal to Capt. A. Kyd to Survey Penang, Jeraja [sic] and Phuket, 11 April 1787: Observations made by Captain Francis Light, 1786," ff. 239–64, Bengal public proceedings, unrecorded papers 726, IOR/G/34/1, BL.

28. Simon Schaffer, "Easily Cracked: Scientific Instruments in States of Disrepair," *Isis* 102:4 (2011): 601–835.

29. Lee T. Macdonald, *Kew Observatory and the Evolution of Victorian Science, 1840–1910* (Pittsburgh, PA: University of Pittsburgh Press, 2018), ch. 2 (ebook version). For examples on the length and complexity of the process, see "Correspondence with Professor Claxton concerning proposals to extend the Hong Kong Observatory, 1911," RGO 7/157(1) Q5, and "Correspondence regarding equipment for the Mount Faber Observatory, Singapore, 1928–1930," RGO 8/93 Q5, CUL.

30. Board's Collections vol. 1906: Draft 670-1841, Collection 15, 13 April 1841, 2, IOR/F/4/1906/81364, and 1 August 1840, 45, IOR/F/4/2026/91387, BL.

31. David Brewster, "XII-On the Mean Temperature of the Equator, as deduced from Observations made at Prince of Wales Island, Singapore and Malacca," *Edinburgh Journal of Science* 8 (1828): 60–67.

32. Henry Atkinson, "XVI—On Astronomical and other Refractions: with a connected Inquiry into the Law of Temperature in different Latitudes and at different Altitudes," *Memoirs of the Astronomical Society of London* 2 (1826): 137–83; Alexander von Humboldt, "XIX—Observations of the Mean Temperature of the Equatorial Regions," *Edinburgh Journal of Science* 6 (1827): 136–44.

33. Anja Werner, "Networks of Knowledge in the Sources of the 1826 Essai politique sur l'île de Cuba," *HiN Alexander von Humboldt im Netz* 15:28 (2014): 59–67.

34. John Bastin and Kwa Chong Guan, *Natural History Drawings: The Complete William Farquhar Collection, Malay Peninsula, 1803–1818* (Singapore: Editions Didier Millet, 2010).

35. J. R. Logan, "The Probable Effects on the Climate of Pinang of the Continued Destruction of Its Hill Jungles," *Journal of the Indian Archipelago* 2 (1848): 534–35.

36. Alfred Russel Wallace, *The Malay Archipelago* (1869) and *The Geographical Distribution of Animals* (1876). See also Martin Krieger, *Nathaniel Wallich: Global Botany in Nineteenth Century India* (London: Routledge, 2023); Martin Fichman, *An Elusive Victorian: The Evolution of Alfred Russel Wallace* (Chicago: Chicago University Press, 2004).

37. "Capt. Low's Dissertation," *Singapore Free Press and Mercantile Advertiser*, 27 October 1836, 2.

38. Robert Little, "Correspondence," *Singapore Free Press and Mercantile Advertiser*, 15 March 1855, 2; "Medical and surgical journal of HMS *Rinaldo* for 1 January to 31 December 1871 by Mr John Buckley, Staff Surgeon, during which time the said ship was employed in the China Station," f. 59r., 101/185/4, Records of the Admiralty, TNA.

39. Charles Burton Buckley, *An Anecdotal History of Old Times in Singapore, 1819–1867* (1902), repr. (Oxford: Oxford University Press, 1984), 685.

40. A. S. Baxendale, "The Cultivation of Cotton in the Federated Malay States," *Agricultural Bulletin of the Straits Settlements and Federated Malay States* 2:11 (November 1903): 345–49.

41. "Cultivation of Cotton in the Federated Malay States," *Straits Times*, 21 December 1903, 2.

42. Paper, "Of an improved thermometer," by James Six, read to the Royal Society, 28 February 1782, L&P/7/240, RS.

Jill Austin and Anita McConnell, eds., *James Six, The Construction of a Thermometer* (London: Nimbus Books, 1980).

43. Letter from James Six to Francis Wollaston, "Variation of local heat as determined by a special thermometer," read to the Royal Society, 10 June 1784, ff. 89r–91v, 94r–v, L&P/8/89, RS.

44. Gerald Mills, "Luke Howard and the Climate of London," *RMetS Weather* 63:6 (2008): 153–54.

45. Luke Howard, *The Climate of London*, 2nd ed. (1833), 3.

46. For modern definitions of scale, see Matthias Roth, "Urban Heat Islands," in *Handbook of Environmental Fluid Dynamics*, ed. Harindra Joseph Shermal Fernando (Boca Raton, FL: CRC Press/Taylor and Francis Group, 2013), 2:144.

47. E. Renou, "Temperatures Differences between City and Country," *Annu-*

aire du Bureau Central Météorologique de France 3:79 (1855; revised 1868); V. Kremser, "The Influence of Large Cities upon Humidity," *Meteorologische Zeitschrift* 25 (1908): 206–15; V. Kremser, "The Results of Wind Measurements Recorded over Many Years in Berlin," *Meteorologische Zeitschrift* 26 (1909): 259–65.

48. N. B. Dennys, *A Descriptive Dictionary of British Malaya* (London: London and China Telegraph Office, 1894), 351; originally written by John Crawfurd in 1856.

49. Dennys, *Descriptive Dictionary*, 352.

50. John Cameron, *Our Tropical Possessions in Malayan India* (London, 1865), 153. See also Walter Makepeace, Gilbert Brooke, and Roland St. John Braddell, eds., *One Hundred Years of Singapore*, 2 vols. (London: John Murray, 1921), 2:282.

51. Alexander C. Maingay, "Report on the Timber and Forest Conservancy of Malacca, 20 August 1865," *SSGG*, 1 September 1865, 388; Allan M. Skinner, "Straits Meteorology," *Journal of the Straits Branch of the Royal Asiatic Society* 11–12 (1883): 252.

52. Skinner, "Straits Meteorology," 248.

53. J. M. Gullick, "A Short History of the Society," *JMBRAS* 68:2 (1995): 67–79, esp. 67.

54. In 1889, for example, the library acquired thirty-right volumes of natural history books. "Annual Report of the Raffles Library and Museum for 1889," *SSGG*, 1 May 1890.

55. Makepeace, Brooke, and Braddell, *One Hundred Years of Singapore*, 2:282.

56. Luke Howard, *The Climate of London deduced from Meteorological Observations*, 1st ed., 3 vols. (London: Harvey and Darton, 1833), 1:2, 10.

57. "Can Man Change Climate?," *Eastern Daily Mail and Straits Morning Advertiser*, 11 January 1907, 6.

58. Untitled article, *Straits Times*, 25 March 1905, 4.

59. "The Recent Heatwave: The Clothing of School Children," *Malaya Tribune*, 18 May 1914, 7.

60. "A Jungle Fire," *Straits Times*, 11 February 1907, 7.

61. Marcus Langdon, *Penang: The Fourth Presidency of India, 1805–1830, Volume Two: Fire, Spice, and Edifice* (Penang: George Town World Heritage, 2015), 175–220. A series of fires in Kampong Glam in the 1840s, including a major one in 1847, are noted in C. Skinner, "Shaer Kampong Gelam Terbakar oleh Abdullah b. Abdul-Kadir, Edited with Notes," *JMBRAS* 45:1 (1972): 21–56, esp. 22–25.

62. Reference from the online Oxford English Dictionary's etymology entry for the noun "heatwave."

63. "The Padre in Scotland," *Singapore Free Press and Mercantile Advertiser*, 2 August 1899, 2.

64. Stephen Sargent Visher, "Local Climates in the Tropics," *Bulletin of the American Meteorological* Society 3:9 (1922): 119–21. An overview of histor-

ic Japanese UHI (urban heat island) studies can be found in Takeshi Sekiguti, "Progress of Climatology in Japan," *Bulletin American Meteorological Society* 35:9 (1954): 428–32. In his summary of urban climate Sekiguti fails to mention some important research by, for example, K. Sasakura, "On the Temperature Difference in and out of Tokyo City," *Geographical Review of Japan* 7 (1931): 551–59; H. Hatakeyama, "The Temperature Distribution in and around Tokyo," *Journal of the Meteorological Society of Japan*, 11 (1932); Hidetoshi Arakawa, "Increasing Air Temperatures in Large Developing Cities," *Beiträge Geophysik* 50 (1937): 3–6; A. Hakoda and T. Kinka, "On Micro-airmass of Osaka and Its Neighbourhood," *Journal of the Meteorological Society of Japan*, 2nd ser., 17 (1939): 425–34; T. Kita, "Town Climate of Tokyo," *Journal of the Meteorological Society of Japan*, 2nd ser. (1939): 155–61.

65. Albert Kratzer, *Climate of Cities*, trans. American Meteorological Society, 2nd ed. (1956), 1–2.

66. Roth, "Urban Heat Islands," 144.

67. James F. Warren, "The Spiral of Failure: Suicide among the Singapore Rickshaw Coolies," *Southeast Asian Journal of Social Science* 13:2 (1985): 47–66, esp. 48.

68. Edward P. Kohn, *Hot Time in the Old Town: The Great Heat Wave of 1896 and the Making of Theodore Roosevelt* (New York: Basic Books, 2010), x.

69. Goh Chor Boon, *Technology and Entrepôt Colonialism in Singapore, 1819–1940* (Singapore: ISEAS, 2013), 105.

70. Colin S. Ruggles, "The Efficient Use of Environmentally-Friendly Natural Rubber Latex in Road Construction—Past, Present and the Future," in *Rubber in Transport Seminar* (Breda: Netherlands, 2004), at http://www.vkrt.org/admin/upload/20041209R.pdf/.

71. Goh Chor Boon, *Technology and Entrepôt Colonialism*, 105.

72. Martin Hendel, Sophie Parison, Arnaud Grados, and Laurent Royon, "Which Pavement Structures Are Best Suited to Limiting the UHI Effect? A Laboratory-Scale Study of Parisian Pavement Structures," *Building and Environment*, 144:15 (2018): 216–29; M. Santamouris, "Using Cool Pavements as a Mitigation Strategy to Fight Urban Heat Island—A Review of the Actual Developments," *Renewable and Sustainable Energy Reviews* 26 (2013): 224–40.

73. Edwin A. Brown, *Indiscreet Memories: 1901 Singapore through the Eyes of a Colonial Englishman* (1935; repr. Singapore: Monsoon Books, 2007), 42.

74. A contrasting example can be found in the following images: "Raffles Place, Singapore c. 1890s," media image no. 19980005492-0036, and "Raffles Place, as Seen from Chartered Bank, 1920," media image no. 19980005492-0035, NAS.

75. "Warm Weather," *Malaya Tribune*, 15 May 1935, 10.

76. "So Hot Last Night," *Nanyang Siang Pau*, 20 June 1938, 30.

77. "Grass Fires," *Malaya Tribune*, 5 February 1935, 13.

78. "Lallang Fires," *Malaya Tribune*, 5 August 1935, 12.

79. "Lallang Blaze at Telok Ayer," *Straits Times*, 24 July 1947, 7.

80. "Two More Lallang Fires in Store," *Straits Times*, 26 July 1947, 3.

81. "Phew! . . . It's Hot!," *Morning Tribune*, 5 April 1948, 1.

82. "Heat Wave Brings Fires in Singapore," *Straits Times*, 6 April 1948, 7.

83. "Our 'Heat Wave,'" *Singapore Free Press and Mercantile Advertiser*, 26 September 1935, 11.

84. "Malaya's Severest Heat Spell," *Singapore Free Press and Mercantile Advertiser*, 8 June 1936, 6.

85. "Malaya's Climate," *Malaya Tribune*, 27 August 1937, 12.

86. Annual averages for Singapore, various locations, collated by the author. For more on these, see Williamson, "Building a Long-Time Series," at https://doi.org/10.5194/cp-17-791-2021/.

87. "Past Climate Trends," at http://www.weather.gov.sg/climate-past-climate-trends/.

88. Alexander W. Still, "Singapore's Future" in Makepeace, Brooke, and Braddell, *One Hundred Years of Singapore*, 2:566.

89. Pan Lu, *Image, Imagination and Imaginarium: Remapping World War II Monuments in Greater China* (Singapore: Palgrave Macmillan, 2021), 134. See also nine photographs of New Kowloon terrain relating to land resumption in the New Territories, Hong Kong: "(6) Photograph of view 'D.' Village of Kau Lung (Kowloon) Tsai and New Kowloon Tong Garden City," CN 3/52, TNA; "Development of Kowloon Tong Estate, 11 December 1929–16 June 1930," CO 129/520/12, TNA. On urban planning to improve lives and livelihoods, see Lucinda Platt, "Poverty Studies and Social Research," in John Holmwood and John Scott, eds., *The Palgrave Handbook of Sociology in Britain* (Basingstoke: Palgrave Macmillan, 2014), 31; Simon Gunn, "Urbanization," in *A Companion to Nineteenth Century Britain*, Blackwell Companions to British History, ed. Chris Williams, 2nd ed. (Maldon, MA: Blackwell Publishing, 2007), 241, 245.

90. Brenda Yeoh, *Contesting Space in Colonial Singapore: Power Relations and the Urban Built Environment* (Singapore: Singapore University Press, 2003), 86–87, 94, 102.

91. S. Nieuwolt, "The Urban Microclimate of Singapore," *Journal of Tropical Geography* 22 (1966): 30–37; Matthias Roth and Winston T. L. Chow, "A Historical Review and Assessment of Urban Heat Island Research in Singapore," *Singapore Journal of Tropical Geography* 33 (2012): 385.

92. "Past Climate Trends," at http://www.weather.gov.sg/climate-past-climate-trends/.

93. N. Nazarian et al., "Integrated Assessment of Urban Overheating Impacts on Human Life," *Earth's Future* 10:8 (2022): e2022EF002682, at doi:10.1029/2022EF002682 (open access).

PART III: CONTROLLING WEATHER

1. Philipp Lehmann, *Desert Edens: Colonial Climate Engineering in an Age of Anxiety* (Princeton, NJ: Princeton University Press, 2022); Meredith McKittrick, "Theories of 'Reprecipitation' and Climate Change in the Settler Colonial World," *History of Meteorology* 8 (2017): 74–94; Yuriko Furuhata, *Transpacific Experiments in Atmospheric Control* (Durham, NC: Duke University Press, 2022).

2. James R. Fleming, *The Checkered History of Weather and Climate Control* (New York: Columbia University Press, 2010); Ruth A. Morgan, *Climate Change and International History: Negotiating Science, Global Change, and Environmental Justice* (London: Bloomsbury Academic Press, 2024), 41–68.

3. S. G. G. Kelliher, "Four Seasons in the Malayan Year," *Straits Times*, 27 June 1936, 10.

4. B. H. F. Barnard, "Effect of Clearing Forest on Rainfall," *Agricultural Bulletin of the Straits and Federated Malay States*, new ser. 10 (1911): 269–70; "Influence of Forests on Drought," *Singapore Free Press and Mercantile Advertiser*, 13 October 1911, 9; "Position of Yunnan," *Straits Times*, 5 July 1920, 9; "Rainmaking," *Singapore Free Press and Mercantile Advertiser*, 31 July 1929, 2; "Bombardment of the Skies," *Singapore Free Press and Mercantile Advertiser*, 22 August 1931, 4.

5. D. H. Wise, "Report on the System of Rice Cultivation in Pahang," *Agricultural Bulletin of the Straits and Federated Malay States* 1:1 (1901): 17.

6. David Henley, "Natural Resource Management and Mismanagement: Observations from Southeast Asian Agricultural History," in *A History of Natural Resources in Asia: The Wealth of Nature*, ed. Greg Bankoff and Peter Boomgaard (London: Palgrave Macmillan, 2007), 22–23; "Soil Erosion in Malaya," *Straits Times*, 20 June 1928, 9.

CHAPTER 5: PLANNING FOR RAINY DAYS

1. Paul Edwards, *A Vast Machine: Computer Models, Climate Data, and the Politics of Global Warming* (Cambridge, MA: MIT Press, 2010), xviii–xix.

2. Rohan D'Souza, "Event, Process and Pulse: Resituating Floods in Environmental Histories of South Asia," *Environment and History* 26:1 (2020): 37. See also Meredith McKittrick, "An Empire of Rivers: The Scheme to Flood the Kalahari, 1919–1945," *Journal of Southern African Studies* 41:3 (2015): 485–504; M. Heffernan, "Bringing the Desert to Bloom: French Ambitions in the Sahara during the Late Nineteenth Century—The Strange Case of la mer intérieure," in *Water, Engineering and Landscape: Water Control and Landscape Transformation in the Modern Period*, ed. D. E. Cosgrove and G. E. Petts (London: Belhaven Press 1990), 94–114; J. M. Powell, *Plains of Promise, Rivers of Destiny: Water Management and the Development of Queensland, 1824–1990* (Trove: Brisbane, 1991).

3. Reyner Banham, *The Architecture of the Well-Tempered Environment* (London: Architecture Press, 1969), 18.

4. D'Souza, "Event, Process and Pulse," 37; Praveen Singh, "Flood Control for North Bihar: An Environmental History from the 'Ground Level,' 1850–1954," in *The British Empire and the Natural World: Environmental Encounters in South Asia*, ed. Deepak Kumar, Vinita Damodaran, and Rohan D'Souza (New Delhi: Oxford University Press, 2011), 160–80.

5. A good overview can be found in Martin Mahony and Georgina Endfield, "Climate and Colonialism," *WIREs Climate Change* 9 (2018): e510. See also Rohan D'Souza, *Drowned and Damned: Colonial Capitalism and Flood Control in Eastern India* (New Delhi: Oxford University Press India, 2016).

6. Cynthia A. Kierner, *Inventing Disaster: The Culture of Calamity from the Jamestown Colony to the Johnstown Flood* (Chapel Hill: University of North Carolina Press, 2019), 2. See also K. Hewitt, ed., *Interpretations of Calamity: From the Viewpoint of Human Ecology* (London: Routledge, 2021); Randolph Kent and John Ratcliffe, *Responding to Catastrophes: U.S. Innovation in a Vulnerable World* (Washington, DC: CSIS Press, 2008).

7. K. J. Fielding, "The Settlement of Penang by James Scott," *JMBRAS* 28:1 (1955): 51.

8. Robert Little, "The Medical Topography of Singapore," *Journal of the Indian Archipelago and Eastern Asia* 3:8 (1848): 469.

9. Marcus Langdon, *Penang: The Fourth Presidency of India, 1805–1830, Vol. Two: Fire, Spice, and Edifice* (Penang: George Town World Heritage, 2015), ch. 1, esp. maps on 45, 149; Gwynn Jenkins, *Contested Space: Cultural Heritage and Identity Reconstructions: Conservations Strategies within a Developing Asian City* (Munster: LIT Verlag, 2008).

10. Wallich is better known as superintendent of the Botanic Gardens at Calcutta. He stayed in Singapore for around three months in 1822 while on medical leave and became acquainted with Raffles because of their mutual interest in botany. R. Hanitsch, N. Wallich, James Lumsdaine, and Francis Salmond, "Letters of Nathaniel Wallich relating to the Establishment of Botanical Gardens in Singapore," *JMBRAS* 42:1 (1969): 145–54, esp. 147.

11. Hanitsch, Wallich, Lumsdaine, and Salmond, "Letters of Nathaniel Wallich," 148.

12. O. J. Dale, *Urban Planning in Singapore: Transformation of a City* (Malaysia: Oxford University Press, 1999), 14. For more on the progress and social impacts of colonial Singapore's land reclamation across time, see Miles A. Powell, "Singapore's Lost Coast: Land Reclamation, National Development and the Erasure of Human and Ecological Communities, 1822–Present," *Environment and History* 27:4 (2021): 635–63.

13. Charles Burton Buckley, *An Anecdotal History of Old Times in Singapore*, 2 vols. (1902; repr. Singapore: Forgotten Books, 1984), 1:75, 83; Stephen Dobbs,

The Singapore River: A Social History, 1819–2002 (Singapore: NUS Press, 2003), 25, 27.

14. Buckley, *Anecdotal History*, 1:75, 82–83, 88–89.

15. Military Secretary to the Governor, 13.10.1829, Singapore Consultations A60, 306–11, Straits Settlements Records, NAS.

16. Despite the epithet "New," the harbor was anything but. Records suggest it had been in use since at least the sixteenth century for commercial activities, because of its deep water and sheltered aspect. C. A. Gibson-Hill, "Singapore Old Strait and New Harbour, 1300–1870," *Memoirs of the Raffles Museum* (Singapore: Government Printing Office, 1956), 3:12–13.

17. Dobbs, *Singapore River*, 11; Gibson-Hill, "Singapore Old Strait," 91.

18. "Municipal Council," *Straits Times*, 26 November 1864, 7; "The Straits Finances," *Straits Times*, 1 April 1865, 2; "Collyer Quay, Singapore c. 1900," media image no. 19980005492-0012, Photographs Collection, NAS.

19. "Collyer of Quay Fame," *Singapore Free Press and Mercantile Advertiser*, 29 June 1933, 8.

20. "Collyer of Quay Fame," 8.

21. Dobbs, *Singapore River*, 11–12.

22. G. M. Reith, *Handbook to Singapore* (1892), revised by Walter Makepeace with an introduction by Paul Kratoska (Singapore: Oxford University Press, 1985), 19. The reclamation (before and after) can be seen clearly on McNair's survey: "Teluk (Telok) Ayer Reclamation c. 1873–1881," media image no. 20050000977, NAS. See also Dale, *Urban Planning*, 18.

23. "Report on the Teluk Ayer Reclamation Scheme," 8 August 1921, RM I.E/55, Straits Settlements Legislative Council Proceedings, 1921, NAS.

24. Robertson quoted in Langdon, *Penang*, 2:531–32.

25. Langdon, *Penang*, 2:531–45.

26. Jarman, Robert L., ed. *Annual Report of the Straits Settlements, Vol. 5, 1894* (Slough: Archive Editions, 1998), 111.

27. Buckley, *Anecdotal History*, 2:626.

28. "Singapore Municipal Committee," *Straits Times*, 8 January 1856, 5.

29. For a comprehensive introduction to the municipal council of Singapore, see Brenda Yeoh, *Contesting Space in Colonial Singapore: Power Relations and the Urban Built Environment* (Singapore: NUS Press, 2003), ch. 2.

30. See, for example, "The Municipality of George Town," *Straits Independent and Penang Chronicle*, 17 April 1895, 5; "Municipality of George Town, Penang: Report for 1907," *Straits Echo*, 10 June 1908, 6; "George Town's Budget," *Straits Echo*, 5 November 1904, 4, "Penang in 1912: Municipal President's Report," *Straits Echo*, 12 July 1913, 8.

31. F. J. Hallifax, "Municipal Government," in *One Hundred Years of Singapore*, ed. Walter Makepeace, Gilbert E. Brooke, and Roland St John Braddell (London: John Murray, 1921), 1:315–40, esp. 317–18.

32. Timothy P. Barnard, "Mosquitos, Public Health and the Construction of a Modern Society," in *Singaporean Creatures: Histories of Humans and Other Animals in the Garden City*, ed. Timothy P. Barnard (Singapore: NUS Press, 2024), 67–99.

33. Buckley, *Anecdotal History*, 1:83.

34. Hanitsch, Wallich, Lumsdaine, and Salmond, "Letters of Nathaniel Wallich," 148.

35. J. Crawfurd, *Journal of an Embassy to the Courts of Siam and Cochin-China* (London, 1828), 2:351.

36. John Caswall, *Observations on the Medical Topography of Singapore* (1830), ff. 2, 6, 22–24, MSS EUR D157, BL.

37. Little, "Medical Topography," 450–52; T. M. Ward and J. P. Grant, "Contributions to the Medical Topography of Prince of Wales Island or Pulo Pinang," *Official Papers on the Medical Statistics and Topography of Malacca and Prince of Wales' Island . . . &c* (Pinang: Government Press, 1830): 1–58, esp. 5–6.

38. Little, "Medical Topography," 458, 465–66.

39. "Pinang," *Singapore Free Press and Mercantile Advertiser*, 3 August 1843, 1.

40. C. C. James, *Drainage Problems of the East* (Bombay: Times of India Office, 1906), 302–3.

41. James, *Drainage Problems*, 302.

42. James, *Drainage Problems*, 303.

43. No item title, RM I.E/53 B107, NAS. Dr. W. R. C. Middleton became the first colonial doctor to attain the new Diploma of Public Health and went on to become the first municipal health officer for Singapore. Makepeace, Brooke, and Braddell, *One Hundred Years of Singapore*, 1:322.

44. No article title, *SSGG*, 6 October 1911, 3; Phua Kai Hong and Mary Lai Lin Wong, "From Colonial Economy to Social Equity: History of Public Health in Malaysia," in *Public Health in Asia and the Pacific: Historical and Comparative Perspectives*, ed. Milton J. Lewis and Kerrie L. MacPherson (London: Routledge, 2007), 170–87, esp. 172.

45. Cited in Z. Baber,

46. "Penang," *Singapore Free Press and Mercantile Advertiser*, 16 December 1847, 2.

47. Makepeace, Brooke, and Braddell, *One Hundred Years of Singapore*, 1:321.

48. Robert L. Jarman, ed., *Annual Reports of the Straits Settlements, 1855–1941*, 12 vols. (Slough: Archive Editions, 1998), 1:427.

49. Major James Low quoted from the *Singapore Free Press* in Goh Chor Boon, *Technology and Entrepôt Colonialism in Singapore, 1819–1940* (Singapore: ISEAS Publishing, 2013), 41n30.

50. Jarman, *Annual Reports*, 1:227; "Pinang," *Singapore Free Press and Mercantile Advertiser*, 12 September 1851, 3; Khoo Salma Nasution, *The Chulia in*

Penang: Patronage and Place-Making around the Kapitan Kling Mosque, 1786–1957 (Penang: Areca Books, 2014), 112.

51. Jarman, *Annual Reports*, 1:88–89.

52. "The Municipal Committee," *Straits Times*, 8 January 1856, 5.

53. Minute Book of the Municipal Commission, 1 July 1881, NA425, NAS; untitled article, *Straits Times*, 21 January 1885, 3.

54. *Straits Times Weekly Issue*, 28 April 1891, 5; *Singapore Free Press and Mercantile Advertiser*, 2 June 1892, 2; *Singapore Free Press and Mercantile Advertiser*, 25 June 1892, 2.

55. Minute Book of the Municipal Council held 8 and 23 December 1884, f. 508, NA425, NAS. For more on this see Dobbs, *Singapore River*.

56. "The Singapore Canal," *Singapore Daily Times*, 18 March 1881, 2.

57. "Venice in Singapore: Floods Follow Yesterday Heavy Rainfall," *Straits Times*, 2 March 1909, 7; "Flood Prevention," *Malaya Tribune*, 18 January 1918, 4; "Rochore Canal and Flooding: Municipal Commissioners' Injunction," *Singapore Free Press and Mercantile Advertiser*, 12 July 1923, 7.

58. "Municipal Commission," *Singapore Free Press and Mercantile Advertiser*, 25 January 1919, 12; "Rochore Canal and Flooding: Municipal Commissioners' Injunction," *Singapore Free Press and Mercantile Advertiser*, 12 July 1923, 7. For a full exposition of the complicated structure of municipal government in the early 1880s, see C. M. Turnbull, *The Straits Settlements 1826–1867: Indian Presidency to Crown Colony* (London: The Athlone Press, 1972), 54–101.

59. Saw Swee-Hock, "Population Trends in Singapore, 1819–1967," *Journal of Southeast Asian History* 10:1 (1969): 36–49, esp. 39.

60. "Legislative Council," *Malaya Tribune*, 16 July 1918, 5.

61. W. Dunman, "The Rural Board," *Straits Times*, 19 January 1928, 10; W. Dunman, "The Rural Board" *Malaya Tribune*, 21 February 1928, 10; "Municipal Commission," *Singapore Free Press and Mercantile Advertiser*, 30 July 1921, 12.

62. "George Town's Budget," *Straits Echo*, 5 November 1904, 4.

63. "Municipality of George Town, Penang: Report for 1907," *Straits Echo*, 2 June 1908, 6.

64. For example, "Town and Country in Malaya," *Straits Times*, 4 June 1934, 12; "Rain Causes Trouble at Penang," *Singapore Free Press and Mercantile Advertiser*, 13 August 1938, 9.

65. Most evidence points to a protracted Na Niña event around the mid 1920s. Joëlle L. Gergis, Karl Braganza, Anthony Fowler, Scott Mooney, and James Risbey, "Reconstructing El Niño–Southern Oscillation (ENSO) from High-Resolution Palaeoarchives," *Journal of Quaternary Science*, 21:7 (2006): 707–22; Robert J. Allan and Rosanne D. D'Arrigo, "'Persistent' ENSO Sequences: How Unusual Was the 1990–1995 El Niño?," *The Holocene* 9:1 (1999): 101–18; W. H. Quinn and V. T. Neal, "The Historical Record of El Nino Events," in *Climate since A.D. 1500*, rev. ed., ed. R. S. Bradley and P. D. Jones (London: Routledge, 1995), 623–48.

66. "St Andrew's Church," at http://standrewschurch.org.my/about-us/history/1924-1960-times-of-testing/.

67. "The Flood," *Straits Times*, 30 December 1926, 9. For more on this flood, see Fiona Williamson, "The Great Flood of 1926: Environmental Change and Disaster Governance in British Malaya," *Journal Ecosystem Health and Sustainability* 2:11 (2016): e01248, special issue on the environmental impact of disasters.

68. "Tanglin Cut Off by Floods," *Singapore Free Press and Mercantile Advertiser*, 7 October 1933, 9; "Heavy Floods in Singapore," *Straits Times*, 18 November 1933, 13.

69. A Ratepayer, "Floods in Singapore," *Straits Times*, 14 January 1928, 10.

70. Anon., "Flood Prevention in Singapore," *Singapore Free Press and Mercantile Advertiser*, 25 December 1931, 8.

71. A Ratepayer, "The Floods," *Straits Times*, 26 May 1932, 6.

72. "Action Demanded," *Singapore Free Press*, 9 January 1934, 8.

73. "The Malayan Rivers," *Straits Times*, 5 May 1928, 2.

74. Report sent from the American Consulate at Singapore to the Secretary of State, Washington, 30 August 1928, ff. 713–19, M712/15, NAS.

75. "Soil Erosion," by Lt. Col. P. McNee, Irrigation Department Kuala Lumpur, 19 November 1945, MU 1963/46, 8a, NAM.

76. "River Improvement in Malaya," *Singapore Free Press and Mercantile Advertiser*, 21 April 1922, 4.

77. "Minimising the Flood Menace," *Straits Times*, 14 January 1934, 11; "Singapore Flood Problem to Be Tackled," *Singapore Free Press and Mercantile Advertiser*, 15 January 1934, 8.

78. "The Governor's Budget Day Speech in Full," *Malaya Tribune*, 31 October 1938, 12.

79. "Chutney," *Straits Maritime Journal and General News*, 19 September 1896, 5; "Heavy Rain at Penang," *Malaya Tribune*, 19 September 1939, 3; "Rain Causes Trouble at Penang," *Singapore Free Press and Mercantile Advertiser*, 13 August 1938, 9.

80. For more on this, see Dobbs, *Singapore River*.

81. For the reservoir delays, see W. H. M. Read, *Play and Politics: Recollections of Malaya* (London: Wells Gardner, Darton, 1901).

82. Khoo Kay Kim, "Developments Relevant to Malayan Agriculture in the Post-Rubber Crisis Era (1920–1921)," *JMBRAS* 72:2 (1999): 17–47, esp. 26, 29.

83. Lynn Hollen Lees, *Planting Empire, Cultivating Subjects: British Malaya, 1786–1941* (Cambridge: Cambridge University Press, 2017), 171, 174.

84. Kenji Koike, "Managing Agency Capitalism and Malayan Rubber: Harrisons & Crosfield, Ltd. (1900–1940)," *JMBRAS* 90:312 (2017): 73–100, esp. 74–76.

85. Koike, "Managing Agency Capitalism," 86.

86. Geoffrey Evans, "The Rubber Research Institute of Malaya," *Nature* 3714 (1941): 15–16.

87. Khoo Kay Kim, "Developments Relevant to Malayan Agriculture," 31–32.

88. Letter to the Secretary to the High Commissioner for the Malay States, Singapore, from O. E. Venables, Under Secretary to Government, Federated Malay States (FMS), 6 January 1939, 3a, Kelantan 817/1938, NAM; Letter to the Secretary to the High Commissioner for the Malay States, Singapore, from M. C. Sheppard, Atg. Under Secretary to Government, FMS, 14 February 1939, 5a, Kelantan 817/1938, NAM; Draft Dispatch to Secretary of State in reply to the S.S. & M.S dispatch from the Secretary of State, 22 August 1939, Kelantan 332/1938, NAM.

89. Paul H. Kratoska, "Imperial Unity versus Local Autonomy: British Malaya and the Depression of the 1930s," in *Weathering the Storm: The Economies of Southeast Asia in the 1930s Depression*, ed. Peter Boomgaard and Ian Brown (Singapore: Institute of Southeast Asian Studies, 2000), 271–94, esp. 271.

90. Kratoska, "Imperial Unity versus Local Autonomy," 282.

91. Phua Kai Hong and Mary Lai Lin Wong, "From Colonial Economy to Social Equity," 172. A full overview of the Health Branch duties can be found in Wellcome Trust Collection P24 J. W. Scharff, *A Note on Public Health Administration in the Northern Settlement (Malaya)*, 1933.

92. W. J. Simpson, *Report on the Sanitary Condition of Singapore* (London: Waterlow, 1907), 26–27.

93. Brenda S. A. Yeoh, *Contesting Space in Colonial Singapore: Power Relations and the Urban Built Environment* (Singapore: NUS Press, 2003): 161–67.

94. Goh Chor Boon, *Technology and Entrepôt Colonialism*, 105; "Heaviest Rainfall in City for 46 Years. Several Districts Flooded: Many Cars Marooned. Office Workers Delayed by First Blow of Monsoon," *Straits Times*, 15 December 1939, 12.

95. For contemporary elevation, see "Elevation of Singapore," at http://elevation.maplogs.com/poi/singapore.64.html/.

96. Untitled article, *Straits Times*, 29 October 1933, 8.

97. James F. Warren, *Rickshaw Coolie: A People's History of Singapore, 1880–1940* (Singapore: NUS Press, 2003), 261–62; Saw Swee-Hock, "Population Trends," 39.

98. Saw Swee-Hock, "Population Trends," 39.

99. Simpson, *Sanitary Condition of Singapore*.

100. Nordin Hussin, *Trade and Society in the Straits of Melaka: Dutch Melaka and English Penang, 1780–1830* (Singapore: NUS Press, 2006), 237–45; "The Deluge," *Straits Echo*, 16 September 1909, 4. For a visualization, see "The Map of Early Penang Showing the Malay Town South of the Town Centre by Sir Home Riggs Popham, 1799" (Wikimedia Commons, public domain).

101. Dobbs, *Singapore River*, 8.

102. Crawfurd, *Journal of an Embassy* 1:66–68, quoted in Gibson-Hill, "Singapore Old Strait," 77.

103. Little, "Medical Topography," 469.

104. Contemporary paintings of such kampong can be found in Wong Hong Suen, *Singapore through 19th Century Prints and Paintings* (Singapore: Editions Didier Millet and National Museum of Singapore, 2010), 132–35.

105. Timothy P. Barnard, "Celates, Raya-Laut, Pirates: The Orang Laut and Their Decline in History," *JMBRAS* 80:2 (2007): 33.

106. Isher Singh Sekhon, "Agriculture in Singapore (1819–1959)," (unpublished bachelor's thesis, University of Malaya, 1961), 23–40, cited in Cynthia Chou, "Agriculture and the End of Framing in Singapore," in Timothy P. Barnard, ed., *Nature Contained: Environmental Histories of Singapore* (Singapore: NUS Press, 2014), 216–40, esp. 224.

107. For more on this event, see Fiona Williamson, "The Politics of Disaster: The Great Singapore Flood of 1954," *Environment and Planning E: Nature and Space* 1:3 (2018): 323–39.

108. Bedok Flood Committee to the Colonial Secretary, 19 January 1955, Singapore Floods 1954–55, PRO15, 671/54, NAS. See also "Stop This Fumbling and DO Get Going, They Tell the Govt," *Straits Times*, 11 December 1954, 10.

109. Gregory Claeys, "Political Thought," in *A Companion to Nineteenth Century Britain: Blackwell Companions to British History*, ed. Chris Williams (Maldon, MA: Blackwell Publishing, 2007), 200.

110. Williamson, "Politics of Disaster."

111. Roland Braddell, *The Lights of Singapore*, 2nd ed. (London: Methuen, 1934), 33.

112. Appointment of Anti-Malaria Committee, 1911, RM I.E/53 B107, NAS.

113. Bas van Bavel, Daniel R. Curtis, Jessica Dijkman, Matthew Hannafor, Maïka de Keyzer, Eline van Onacker, and Tim Soens, eds., *Disasters and History: The Vulnerability and Resilience of Past Societies* (Cambridge: Cambridge University Press, 2020), 9.

CHAPTER 6: DEALING WITH DROUGHT

1. Sunil Amrith, *Unruly Waters: How Rains, Coasts, and Seas have Shaped Asia's History* (London: Penguin, 2018); David Gilmartin, "Models of the Hydraulic Environment: Colonial Irrigation, State Power and Community in the Indus Basin," in David Arnold and Ramchandra Guha, eds., *Nature, Culture, Imperialism: Essays on the Environmental History of South Asia* (Oxford: Oxford University Press, 1995), 76–103.

2. Raghav Kishore, "Urban 'Failures': Municipal Governance, Planning and Power in Colonial Delhi, 1863–1910," *Indian Economic and Social History Review* 52:4 (2015): 439–61; Gilmartin, "Models of the Hydraulic Environment"; David Gilmartin, "Scientific Empire and Imperial Science: Colonialism and Irrigation Technology in the Indus Basin," *Journal of Asian Studies* 53:4 (1994): 1127–49.

See also Timothy Mitchell, *Rule of Experts: Egypt, Techno-Politics, Modernity* (Berkeley: University of California Press, 2002).

3. Stephen Dobbs, *The Singapore River: A Social History, 1819–2002* (Singapore: NUS Press, 2003), 33–34.

4. *Singapore Free Press and Mercantile Advertiser*, 14 February 1856, 3; W. H. M. Read, *Play and Politics: Recollections of Malaya* (London: Wells Gardner, Darton, 1901); Brenda Yeoh, "Urban Sanitation, Health and Water Supply in Late Nineteenth and Early Twentieth Century Colonial Singapore," *Southeast Asia Research* 1:2 (1993): 146–48.

5. John Broich, "Engineering the Empire: British Water Supply Systems and Colonial Societies, 1850–1900," *Journal of British Studies* 46:2 (2007): 346–65.

6. Maintenance bills started in November 1877. See *Straits Settlements Government Gazette* (hereafter *SSGG*), 26 December 1879, 2028.

7. Nordin Hussin, *Trade and Society in the Straits of Melaka: Dutch Melaka and English Penang, 1780–1830* (Singapore: NUS Press, 2007), 251.

8. T. J. Newbold, *Political and Statistical Account of the British Settlements in the Straits of Malacca* (1839), vol. 1; *Singapore Chronicle and Commercial Register*, 24 March 1831, 1; *SSGG*, 13 July 1886, 1128; Robert Jarman, ed., *Annual Reports of the Straits Settlements, 1855–1941*, 12 vols. (Slough: Archive Editions, 1998), vol. 4.

9. Brenda Yeoh, *Contesting Space in Colonial Singapore: Power Relations and the Urban Built Environment* (Singapore: NUS Press, 2003), 175–212; Yeoh, "Urban Sanitation."

10. Evidence compiled by the author from diverse sources, including newspapers, government gazettes, and antiquarian accounts and memoirs. The severe 1877–1878 event was likely a protracted El Niño. Robert J. Allan, Joëlle Gergis, and Rosanne D'Arrigo, "Placing the AD 2014–2016 'protracted' El Niño into a Long-Term Context," *The Holocene* 30:1 (2020): 103; Joëlle L. Gergis and A. M. Fowler, "A History of ENSO Events since AD 1525: Implications for Future Climate Change," *Climatic Change* 92 (2009): 375.

11. H. W. Jack, "Rice in Malaya," *Malayan Agricultural Journal* 11:5 (1923), 103; R. D. Hill, *Rice in Malaya: A Study in Historical Geography* (Singapore: NUS Press, 2012), 31–32, 34–37.

12. *SSGG*, July to September 1898, appendix "Imports 1897," 19.

13. Nathaniel Cantley, "Report on the Forests of the Straits Settlements, 1882," *Annual Report on the Botanic Gardens, Singapore, 1884* (Singapore, 1885), 1–7.

14. John K. Corner, *My Father in His Suitcase: In Search of E. J. H. Corner, the Relentless Botanist* (Singapore: Landmark Books, 2013), 80, 91.

15. "Rainfall and Deforestation," *Singapore Free Press and Mercantile Advertiser*, 15 June 1905, 371.

16. Helen Tilley, *Africa as a Living Laboratory: Empire, Development, and the Problem of Scientific Knowledge, 1870–1950* (Chicago: University of Chicago

Press, 2011), 21; Christian Jennings, "Unexploited Assets: Imperial Imagination, Practical Limitations, and Marine Fisheries Research in East Africa, 1917–53," in *Science and Empire: Knowledge and Networks of Science across the British Empire, 1800–1970*, ed. Brett M. Bennett and Joseph M. Hodge (Basingstoke: Palgrave Macmillan, 2011), ebook.

17. Timothy P. Barnard and Joanna W. C. Lee, "A Spiteful Campaign: Agriculture, Forests, and Administering the Environment in Imperial Singapore," *Environmental History* 27:3 (2022): 467, 473.

18. Jeyamalar Kathirithamby-Wells, *Nature and Nation: Forests and Development in Peninsular Malaysia* (Singapore: NUS Press, 2005), 74, 88; A. M. Burns-Murdoch, "Forest Administration Report of the Federated Malay States, 1906," *Indian Forester* 33 (1907): 430.

19. Barnard and Lee, "Spiteful Campaign," 475–76.

20. Lim Teck Ghee, *Peasant Agriculture in Colonial Malaya, 1874–1941* (PhD thesis, Australian National University, 1971), 141.

21. Francis S. P. Ng, *100 Years of Tropical Forest Research: The Story of the Forest Research Institute Malaysia* (Selangor: Forest Research Institute Malaysia, 2010), 6.

22. Ng, *100 Years*, 13.

23. A. M. Burns-Murdoch, *The Trees and Timbers of the Malay Peninsula* (1911), preface.

24. K. M. Wong, "The Herbarium and Arboretum of the Forest Research Institute Malaysia at Kepong—a Historical Perspective," *Garden's Bulletin Singapore* 40:1 (1987): 16; K. D. Menon, "A Brief History of Forest Research in Malaya," *Commonwealth Forestry Review* 48:4 (1969): 309–16.

25. Note from James Watson to Foxworthy dated 23 September 1928, quoted from Wong, "Herbarium and Arboretum," 16.

26. C. B. Buckley, *An Anecdotal History of Old Times in Singapore, 1819–1867*, repr. (1902; Oxford: Oxford University Press, 1984), 685.

27. John Cameron, *Our Tropical Possessions in Malayan India* (London: John Murray, 1865), 150.

28. Victor Lieberman and Brendan Buckley, "The Impact of Climate on Southeast Asia, c. 950–1820: New Findings," *Modern Asian Studies* 46:5 (2012): 1050, 1054, 1091, 1057–58.

29. "Remarks on the Climate, Soil and Cultivation of Penang and Province Wellesley," *Singapore Chronicle and Commercial Register*, 4 September 1834, 1.

30. "Meteorological Observations," *Singapore Free Press and Mercantile Advertiser*, 6 April 1837, 3.

31. A Correspondent, "Capt. Low's Dissertation," *Singapore Free Press and Mercantile Advertiser*, 20 October 1836, 2. See also James Low, *A Dissertation on the Soil & Agriculture of the British Settlement of Penang, Or Prince of Wales Island, in the Straits of Malacca: Including Province Wellesley on the Malayan Peninsula.*

With Brief References to the Settlements of Singapore & Malacca (Singapore: Singapore Free Press Office, 1836), 817–18.

32. Untitled article, *Straits Times*, 1 May 1875, 5; "Drought at Penang," *Straits Times*, 24 April 1899, 2.

33. Cecilia Tortajada, Yual Joshi, and Asit K. Biswas, *The Singapore Water Story: Sustainable Development in an Urban City-State* (London: Routledge, 2013), 8.

34. F. J. Halifax, "Municipal Government," in Walter Makepeace, Gilbert Brooke, Roland St. John Braddell, eds., *One Hundred Years of Singapore* (London: John Murray, 1921), 1:316.

35. Hussin, *Trade and Society*, ch. 4.

36. Amitav Ghosh, *The Nutmeg's Curse: Parables for a Planet in Crisis* (Chicago: University of Chicago Press, 2021), 108–9, 178–79, Amrith, *Unruly Waters*, ebook.

37. Bas van Bavel, Daniel R. Curtis, Jessica Dijkman, Matthew Hannaford, Maïka de Keyzer, Eline van Onacker, Tim Soens, eds., *Disasters and History: The Vulnerability and Resilience of Past Societies* (Cambridge: Cambridge University Press, 2020), 10–11, 74.

38. Saw Swee-Hock, "Population Trends in Singapore, 1819–1967," *Journal of Southeast Asian History* 10:1 (1969): 37.

39. Jarman, *Annual Reports of the Straits Settlements*, 8:41, 181, 311.

40. W. J. Simpson, *Report on the Sanitary Condition of Singapore* (London: Waterlow, 1907), 6.

41. Ng Siew Yoong, "The Chinese Protectorate in Singapore, 1877–1900," *Journal of Southeast Asian History* 2:1 (1961): 81–83.

42. James F. Warren, "The Spiral of Failure: Suicide among the Singapore Rickshaw Coolies," *Southeast Asian Journal of Social Science* 13:2 (1985): 48.

43. Ben Wisner, Piers Blaikie, Terry Cannon and Ian Davies, eds., *At Risk: Natural Hazards, People's Vulnerability and Disasters*, 2nd ed. (London: Routledge, 2004).

44. Warren, "Spiral of Failure" 47.

45. Ng Siew Yoong, "Chinese Protectorate," 94; Leon Comber, "Chinese Secret Societies in Malaya: An Introduction," *Journal of the Malayan Branch of the Royal Asiatic Society* 29:1 (1956): 149.

46. Comber, "Chinese Secret Societies," 149, 153.

47. Warren, "Spiral of Failure," 50.

48. "Price of Rice," *Singapore Free Press and Mercantile Advertiser*, 11 September 1911, 4.

49. James F. Warren, *Rickshaw Coolie: A People's History of Singapore, 1880–1940* (Singapore: NUS Press, 2003), 5, 261.

50. Yeoh, *Contesting Space*, 207.

51. Yeoh, *Contesting Space*, 202.

52. Yeoh, "Urban Sanitation," 149.

53. Tan Yong Soon, Lee Tung Jean, and Karen Tan, *Clean, Green, and Blue: Singapore's Journey towards Environmental and Water Sustainability* (Singapore: ISEA Publishing, 2009), 183; Tortajada, Joshi, and Biswas, *Singapore Water Story*, 45.

54. Van Bavel et al., *Disasters and History*, 78.

55. Halifax, "Municipal Government," 328.

56. "Effects of the Weather," *Singapore Free Press and Mercantile Advertiser*, 11 March 1911, 7; untitled article, *Straits Times*, 24 March 1911, 6.

57. "Insanitary Singapore," *Weekly Sun*, 3 June 1911, 6.

58. *SSGG*, 26 May 1911, 792; *SSGG*, 16 June 1911, 872.

59. Brittany Merritt, "Insecurities of Empire: Struggles over Health Reform in Interwar Barbados," *Journal of Colonialism and Colonial History* 18:3 (2017): 1–13; Peter Baldwin, *Contagion and the State in Europe, 1830–1930* (Cambridge: Cambridge University Press, 2004).

60. P. N. Gerrard, *The Hygienic Management of Labour in the Tropics* (Singapore: Methodist Publishing House, 1913), 12–14.

61. Yeoh, "Urban Sanitation," 146.

62. Untitled article, *Straits Times*, 5 May 1877, 3.

63. "Water Famine," *Straits Times*, 3 September 1902, 5; Untitled article, *Straits Times*, 4 September 1902, 4.

64. "Report of the Municipal Health Officer," *SSGG*, August 1896, 9.

65. Cited in Warren, *Rickshaw Coolie*, 261.

66. Untitled article, *Straits Times*, 29 August 1902, 4; untitled article, *Straits Times*, 30 August 1902, 4.

67. "Water Famine," *Straits Times*, 3 September 1902, 5.

68. "The Water Supply," *Straits Times*, 13 September 1902, 5.

69. "Health Department Report," *SSGG* (26 May 1911): 792.

70. "Appendix to the Annual Health Report for Singapore, 1921," *Report of the Analytical Laboratory*, 10 April 1922, 35–38-E.

71. Warren, *Rickshaw Coolie*, 258–60.

72. James Beattie and Ruth Morgan, "Engineering Edens on This 'Rivered Earth'? A Review Article on Water Management and Hydro-Resilience in the British Empire, 1860s–1940s," *Environment and History* 23 (2017): 50.

73. Mark Elvin, *The Retreat of the Elephants: An Environmental History of China* (New Haven, CT: Yale University Press, 2004).

74. Broich, "Engineering the Empire," 359–60; Roy M. McLeod, "Scientific Advice for British India: Imperial Perceptions and Administrative Goals, 1898–1923," *Modern Asian Studies* 9:3 (1975): 343–84.

75. van Bavel et al., *Disasters and History*, 113–14.

76. "Singapore's Water," *Straits Times*, 27 March 1912, 7.

77. "Bridge or Causeway across Johore Straits," Despatches High Commission and Offices for the Straits Settlements and Malay States, no. 402, 97–125, CO273/447, TNA.

78. G. Alphonso and Albert Lau, *The Causeway* (Kuala Lumpur and Singapore: National Archives of Malaysia and National Archives of Singapore, 2011), 92–100.

79. "Singapore Municipal Finances," *Straits Times*, 25 August 1930, 6; A. Rahman Tang, "Revisiting the Water Issue across the Causeway: The Origins of Water Supply from Johor to Singapore, 1904–32," *Jebat: Malaysian Journal of History, Politics and Strategy* 40:1 (2013): 1–20.

80. "Municipal Impressions," *Straits Times*, 1 June 1928, 12.

81. "Singapore Municipal Finances," *Straits Times*, 25 August 1930, 6; "Report of the Commission on the Financial Position of the Municipality," part 6, C5, Straits Settlements Legislative Council, 23 January 1922, CO275/106, TNA; *SSGG*, supplement, 4 August 1922, 8–9, 30 September 1922, 3.

82. "Gunong Pulai Water," *Straits Times*, 18 February 1927, 9; *SSGG*, 1 April 1927, 1; *SSGG*, 10 June 1927, 1.

83. Report of the Commission on the Financial Position of the Municipality, part 6, C5, Straits Settlements Legislative Council, 23 January 1922, CO275/106, TNA.

84. Report of the Commission on the Financial Position of the Municipality, Appendix E, C27, Report of the Commission on the Financial Position of the Municipality, Diagram comparing Water Resources, Straits Settlements Legislative Council, 23 January 1922, CO275/106, TNA.

85. "Diagram comparing Water Resources." C30–31, TNA.

86. Frédéric Graber, "Inventing Needs: Expertise and Water Supply in Late Eighteenth and Early Nineteenth-Century Paris," *British Journal for the History of Science* 40:3 (2007): 316–17.

87. Murnane quoted from "Long View," *Straits Times*, 26 June 1938, 17.

88. "Mr David Murnane," *Singapore Free Press*, 30 January 1953, 5.

89. Christopher Bayly and Tim Harper, *Forgotten Armies: Britain's Asian Empire and the War with Japan* (Cambridge, MA: Belknap Press, 2005), 116.

90. Margaret Shennan, *Out in the Midday Sun: The British in Malaya, 1880–1960* (Singapore: Monsoon Books, 2015), 322; "Penang Bombing," *Singapore Free Press and Mercantile Advertiser*, 13 December 1941, 5; "Governor on Evacuation of Penang," *Singapore Free Press and Mercantile Advertiser*, 23 December 1941, 1; "Prai Power Station Now Nearly Ready," *Syonan Shimbun*, 4 December 1942, 2.

91. Mare Senki Shingeki No Kiroku (4 of 8), 1942, Audio-Visual and Sound Recordings, 2011005477, NAS; Kevin Blackburn and Karl Hack, *Did Singapore Have to Fall? Churchill and the Impregnable Fortress* (London: Routledge, 2004), 122; Ronald McCrum, *Men Who Lost Singapore, 1938–1942* (Singapore: NUS Press, 2017), ebook.

92. Hansard: Navy Supplementary Estimate 1927, vol. 215, debated in the Commons Chamber on 22 March 1928: Comments of Mr Mardy Jones, Column 692.

93. Paul Kratoska, *The Japanese Occupation of Malaya and Singapore, 1941–45*, 2nd ed. (Singapore: NUS Press, 2018), 42–43.

94. Blackburn and Hack, *Did Singapore Have to Fall?*, 117.

95. Untitled article, *Straits Times*, 16 April 1926, 8.

96. "Notes of the Day: Sunshine without Thirst," *Straits Times*, 21 January 1930, 12.

97. "1930—Singapore's Driest Year?," *Straits Times*, 25 September 1930, 12.

98. Yeoh, "Urban Sanitation," 161–62, 166.

99. "Singapore's $5,500,000 Water Scheme," *Straits Times*, 2 April 1939, 17.

100. Notes on correspondence with the Treasury by A. N. Galsworthy, 6 October 1939, f. 9, 50639, CO273/659, TNA.

101. Letter from the Treasury Chambers to the Colonial Office, 2 November 1939, f. 21, 50639, CO273/659, TNA.

102. "Penang Waterworks: Special Expenditure," Annual Report of the Public Works Department for 1939, 9, Straits Settlements Annual Reports 1939–40, R. M. I D/83, NAS.

103. "For Health, Victory Grow Vegetables," *Malaya Tribune*, 12 April 1940, 2.

104. "Drought Breaks in the F.M.S.," *Straits Budget*, 5 October 1939, 5.

105. "Severe Drought at Kuala Kangsar," *Straits Budget*, 7 September 1939, 6.

106. "Malacca's Long Drought," *Straits Times*, 10 March 1940, 9.

107. "Drought Threatens Padi-Growers," *Straits Budget* 27 July 1939, 13.

108. "Estates Obtain Water from Town by Lorry," *Straits Budget*, 21 March 1940, 11; "Drought in North Kedah," *Pinang Gazette and Straits Chronicle*, 9 April, 1940, 7; "Scarcity of Rain in N. Johore," *Pinang Gazette and Straits Chronicle*, 30 July 1940, 4.

109. "Severe Drought Sweeps over Alor Gajah," *Malaya Tribune*, 14 July 1939, 14.

110. "Negri Sembilan Needs More Water Schemes," *Straits Budget*, 11 July 1940, 7.

111. Annual Report of the Audit of Accounts of the Straits Settlements for 1940: Expenditure Accounts, 10, Straits Settlements Annual Reports 1939–40, R. M. I D/83, NAS.

112. "Estates Obtain Water from Town by Lorry," *Straits Budget*, 21 March 1940, 11; "Severe Drought Sweeps over Alor Gajah," *Malaya Tribune*, 14 July 1939, 14.

113. "Increasing Singapore's Food Supplies," *Straits Times*, 20 August 1939, 14; "Drought Postpones Agricultural Schemes," *Malaya Tribune*, 22 April 1940, 3.

114. Ang Seah San, "Japanese Occupation of Singapore," accession no. 000419, reel 5, transcript, p. 62, Oral Histories, NAS.

115. To plan for public works, utilities, and infrastructure, for example, they used Singapore Municipality records, which were held in London, under the firm

Peirce and Williams. Interestingly, this was the same Robert Peirce, municipal engineer of Penang and later Singapore, who had returned to Britain after his retirement in 1906.

116. Oswald Gilmour, *With Freedom to Singapore: A Social and Administrative Assessment of Post-War Singapore, 1945–1950* (London: Ernest Benn, 1950), reprinted 2020, ebook.

117. Gilmour, *With Freedom to Singapore*, ebook.

118. "Report of the Selangor Medical Department for the years 1941–46," RC Sel 296/1947, reproduced from Kratoska, *Japanese Occupation*, 330.

119. "Water May Be Rationed in Singapore," *Straits Times*, 10 January 1947, 3.

120. In Britain, the military had undertaken localized weather control experiments to clear fog from British airfields known as the Fog Investigation and Dispersal Operation (FIDO) during the Second World War and shifted into cloud seeding during the 1950s with the divisive "Project Cumulus." For the development of the FIDO scheme, see James R. Fleming, *The Callendar Effect: The Life and Work of Guy Stewart Callendar, 1898–1964* (Boston, MA: American Meteorological Association, 2007), 51–60. For Project Cumulus, see H. C. Shellard and D. R. Grant, "Results of an Artificial Nucleation Experiment," *Meteorological Magazine* 80:951 (1951): 253–55; Meteorological Research Flight: Project Cumulus flight 7: ascent over Cranfield: flight reports, 1952, BJ4, TNA.

121. Benjamin Weil, "The Rivers Come: Colonial Flood Control and Knowledge Systems in the Indus Basin, 1840s–1930s," *Environment and History* 12:1 (2006): 3–29.

122. But see Allan M. Skinner, "Straits Meteorology," *Journal of the Straits Branch of the Royal Asiatic Society* 11–12 (1883): 245–59.

123. Ruth Morgan, "Prophecy and Prediction: Forecasting Drought and Famine in British India and the Australian Colonies," *Global Environment* 13 (2020): 96–133. Discussion on the rivalry between Henry Chamberlain Russell and Charles Egeson on late-nineteenth-century cyclical prediction on 120–24 is especially pertinent here.

124. Henry F. Blanford, "The Eleven-Year Periodical Fluctuation of the Carnatic Rainfall," *Nature* (1877): 227–29; Henry F. Blanford, *Report on the Variation of Rainfall in Tropical India with the cycle of sun-spot activity* (Calcutta, 1877), IOR/V/27/440/9, BL.

125. Morgan, "Prophecy and Prediction," 104, 106, 108; Mike Davis, *Late Victorian Holocausts: El Niño Famines and the Making of the Third World* (London: Verso, 2001), ebook.

126. Norman J. Lockyer, "The Meteorology of the Future," *Nature* (1872): 98.

127. John Eliot, *Report on the Meteorology of India in 1889* (Calcutta: Office of the Superintendent of Government Printing, 1891), 15:283–84. Blanford's book, written in the nineteenth century, was probably the most comprehensive on the subject for the region: Henry F. Blanford, *A Practical Guide to the Climates and*

Weather of India, Ceylon and Burmah, and the Storms of the Indian Seas (London: Macmillan, 1889).

128. C. D. Stewart, *Weather Forecasting in Malaya*, 1935, 1, SEL: SEC G.438/1935, NAM.

129. Stewart, *Weather Forecasting*, 2–3.

130. F. E. A. F.—Establishment of Meteorological Flights, 6 November 1944, TNA, BJ5/173; Fleet Meteorology, 1931, CO825/9/3, TNA.

131. Synoptic Conditions in Far East—Investigation, 7 July 1944, BJ5/157, TNA; Air Ministry Meteorological Research Committee, Minutes of the 47th Meeting, 11 April 1947, BJ5 93, TNA.

132. Malayan Meteorological Service Annual Report, 1948, 7, FS 13173, 1949, NAM.

133. Lee Onn Pong, "The Public Service," accession no. 002510, reel 49, Oral Histories, NAS; "Drought Hits 2,500 in New Village," *Straits Times*, 22 July 1962, 4; "Disease and Drought Hit Rice," *Straits Budget*, 24 October 1962, 11.

134. "Drought in Singapore Causes Water Shortage, 1963," British Pathé/Reuters Film ID: VLVA2QO4NSL8T7N01J8L7PVS0QAA, available online at https://www.britishpathe.com/asset/163771/.

135. B. White, *Report on the Water Resources of Singapore Island, Excluding Those within the Present Protected Catchment Areas* (London: Wolfe Barry and Partners, 1950). It is possible that this report may have been based on an earlier geological report on Singapore, written by Dr. Elizabeth Alexander, wife of Dr. Norman Stanley Alexander, "Cavendish Laboratories. And Government Appointed Geologist to Singapore, during the 1940s," Lim Ho Hup, "The Public Service," accession no. 002247, reel 6, transcript, p. 71, and reel 7, transcript, p. 73, Oral Histories, NAS. Certainly, her survey was used to underpin the later study of water resources in Singapore in 1960s undertaken by a hydrologist, Lim Ho Hup, during his time at Singapore's Environmental Development Board. See reel 6, transcript, p. 71, Oral Histories, NAS.

136. Lee Kuan Yew, interview by Biswas and Tortajada, 2009, retrieved from the Centre for Liveable Cities and Public Utilities Board Singapore's Urban Systems Studies Booklet Series: *Water: From Scarce Resource to National Asset* (Cengage Learning Asia, 2012), 3–4.

CHAPTER 7: REGULATING HEAT, CONTROLLING URBAN AIRS

1. W. J. Simpson, *Report on the Sanitary Condition of Singapore* (London: Waterlow, 1907), 5.

2. Philipp Lehmann, *Desert Edens: Colonial Climate Engineering in the Age of Anxiety* (Princeton, NJ: Princeton University Press, 2022); Kristine C. Harper, *Make It Rain: State Control of the Atmosphere in Twentieth-Century America* (Chicago: University of Chicago Press, 2017); James R. Fleming, *Fixing the Sky: The*

Chequered History of Weather and Climate Control (New York: Columbia University Press, 2010); James R. Fleming, "The Climate Engineers," *Wilson Quarterly* 31:2 (2007): 51.

3. Michael Thad Allen and Gabrielle Hecht, "Authority, Political Machines, and Technology's History," in *Technologies of Power: Essays in Honour of Thomas Parke Hughes and Agatha Chipley Hughes*, ed. Michael Thad Allen and Gabrielle Hecht (Cambridge, MA: MIT Press, 2001), 1–24.

4. On air-conditioning from architectural or popular culture perspectives, see Joseph M. Siry, *Air-Conditioning in Modern American Architecture, 1890–1970* (University Park, PA: Pennsylvania State University Press, 2022); Daniel A. Barber, *Modern Architecture and Climate: Design before Air-Conditioning* (Princeton, NJ: Princeton University Press, 2020); Marsha E. Ackernmann, *Cool Comfort: America's Romance with Air-Conditioning* (Washington, DC: Smithsonian Institution Press, 2002); John E. Crowley, *The Invention of Comfort: Sensibilities and Design in Early Modern Britain and Early America* (Baltimore: Johns Hopkins University Press, 2001).

5. Chris Courtney, "The Birth of Cool: Heat and Air-Conditioning in the History of Wuhan, 1950–2020," *Urban Studies* (2023): 1–16, accessed August 23, 2023, at doi.org/10.1177/00420980231153512/; David Arnold, "Europe, Technology, and Colonialism in the 20th Century," *History and Technology* 21:1 (2005): 85–106.

6. Kah-Wee Lee and Lawrence Chua, "Modernity's Other," *South East Asia Research* 28:2 (2020): 112.

7. Goh Chor Boon, "Imported Technology: Its Idea and Development," *JMBRAS* 71:1 (1998): 45–46, 48–49. See also Philip Smith, "Where There Is Life: Science Fiction in Singapore," *Journal of Commonwealth Literature* 57:2 (2019): 296. Smith quotes from R. Patke and P. Holden, *The Routledge Concise History of Southeast Asian Writing* (New York: Routledge, 2010), 93.

8. K. G. Tregonning, "Tertiary Education in Malaya: Policy and Practice, 1905–1962," *JMBRAS* 63:1 (1990): 2; Erman Agaoglu and Abu Hanifah Haris, "Technical Education in Malaya: The Development of Technical School Kuala Lumpur, 1925–1940," *Sejarah: Journal of History Department, University of Malaya* 29:2 (2020): 111.

9. Sheila Jasanoff, "Future Imperfect: Science, Technology, and the Imaginations of Modernity," in *Sociotechnical Imaginaries and the Fabrication of Power*, ed. Sheila Jasanoff and Sang-Hyun Kim (Chicago: University of Chicago Press, 2015), 19.

10. Arnold, "Europe, Technology, and Colonialism," 90.

11. For a good introduction, see Jiat Hwee Chang and Anthony D. King, "Towards a Genealogy of Tropical Architecture: Historical Fragments of Power-Knowledge, Built Environment and Climate in the British Colonial Territories," *Singapore Journal of Tropical Geography* 32 (2011): 283–300.

12. Wong Hong Suen, *Singapore through 19th Century Prints and Paintings* (Singapore: Editions Didier Millet and National Museum of Singapore, 2010), 134–35.

13. Tan Sri Mubin Sheppard, "Traditional Malay House Forms in Trengganu and Kelantan," *JMBRAS* 42:2 (1969): 2.

14. Azizi Bahauddin and Teh Boon Soon, "The Environment and *Feng Shui* Application in Cheong Fatt Tze Mansion, Penang, Malaysia," in *WIT Transactions on The Built Environment* (Ashurst: WIT Press, 2019), 183:1–13.

15. Gwynn Jenkins, *Cultural Heritage and Identity Reconstructions: Conservations Strategies within a Developing Asian City* (Zurich: Lit Verlag, 2008), 235.

16. Andrew Ballantyne and Andrew Law, "Genealogy of the Singaporean Black-and-White House," *Singapore Journal of Tropical Geography* 32 (2011): 301–2; Chang and King, "Towards a Genealogy," 285–87; B. Shelton, J. Karakiewicz and T. Kvan, *The Making of Hong Kong: From Vertical to Volumetric* (London: Routledge, 2010), 31–39; John Weiler, "Colonial Connections: Royal Engineers and Building Technology Transfer in the Nineteenth Century," *Construction History* 12 (1996): 16.

17. T. Roger Smith, "On Buildings for European Occupation," Papers read at The Royal Institute of British Architects 1867–8 (London, 1868): 198, 200. See also, T. Roger Smith, "Architectural Art in India," *Journal of the Society of Arts* (1873): 282.

18. Goh Chor Boon, *Technology and Entrepôt Colonialism in Singapore, 1819–1940* (Singapore: ISEAS Publishing, 2013), 44.

19. All these sites are visible on the 1913 Map of Singapore showing the "Principle Residences and Places of Interest," media image no. 20120001590-0006, NAS. A digital copy can be viewed at the NAS website at https://www.nas.gov.sg/archivesonline/maps_building_plans/record-details/fac6ffc2-115c-11e3-83d5-0050568939ad/.

20. Marjorie Binnie, quoted from Margaret Shennan, *Out in the Midday Sun: The British in Malaya, 1880–1960* (Singapore: Monsoon Books, 2015), 86.

21. John Cameron, *Our Tropical Possessions in Malayan India &c* (Smith Elder and Co., 1865), 158.

22. "Prickly Heat," *Singapore Free Press and Mercantile Advertiser*, 10 July 1889, 43; "The Prevention of Sunstroke," *Singapore Free Press and Mercantile Advertiser*, 18 July 1896, 2.

23. Claire Lowrie, Nicki Tarulevicz, Fiona Williamson, and Charmaine Lam, "Cosmopolitan Elites and the Management of Tropical Heat in Colonial Malaya, c. 1870s–1930s," *Journal of Colonial History and Colonialism* 25:1 (2024): 10.1353/cch.2024.a928069/.

24. Shennan, *Out in the Midday Sun*, 172.

25. Nandini Bhattacharya, *Contagion and Enclaves: Tropical Medicine in Colonial India* (Liverpool: Liverpool University Press, 2012).

26. Robert Aiken, "Early Penang Hill Station," *Geographical Review* 77:4 (1987): 422–25.

27. "Recollections of Malaya: "'Recollections of Thirty-Five Years as a Missionary Teacher in Malaya, by the Reverend Peach, 1962,'" f. 10, RCS 103/4/26, CUL.

28. Reverend Peach, "Recollections of Malaya," f. 9.

29. Lowrie, et al., "Equatorial Living," ff. 1r, 3r, file 10, 1927, CO273/541/4, TNA.

30. Cherian George, *Singapore: The Air-Conditioned Nation: Essays on the Politics of Comfort and Control, 1990–2000* (Singapore: Landmark Books, 2000), 17–18.

31. Robert Jarman, ed., *Annual Reports of the Straits Settlements, 1855–1941*, 12 vols. (Slough: Archive Editions, 1998), 2:203, 586.

32. Timothy Pwee, "Let There Be Light," *BiblioAsia* 16:4 (2021): 36–41.

33. Walter Makepeace, Gilbert E. Brooke, and Roland St. John Braddell, eds., *One Hundred Years of Singapore* (London: John Murray, 1921). 2:596.

34. "The Dennys-Cuff System of Electric Lighting," *Straits Times Weekly Issue*, 15 December 1883, 2.

35. "The Dennys-Cuff System of Electric Lighting," *Straits Times*, 19 December 1883, 9; "New Patents," *Telegraphic Journal and Electrical Review* 14:331 (29 March 1884): 274.

36. Makepeace, Brooke, and Braddell, *One Hundred Years*, 2:214–15.

37. Makepeace, Brooke, and Braddell, *One Hundred Years*, 2:6.

38. "The First Installation of the Electric Light," *Straits Times*, 3 April 1886, 2.

39. "Electric Lighting in Penang," *Straits Times*, 14 September 1900, 2; *Singapore Free Press and Mercantile Advertiser*, 1 November 1900, 2.

40. "Partners in Adventure," *Straits Times*, 25 October 1958, 1; Makepeace, Brooke, and Braddell, *One Hundred Years*, 2:199–200, 600–601, 606.

41. Makepeace, Brooke, and Braddell, *One Hundred Years*, 2:170–71.

42. Pwee, "Let There Be Light."

43. Makepeace, Brooke, and Braddell, *One Hundred Years*, 2:170.

44. Makepeace, Brooke, and Braddell, *One Hundred Years*, 2:318, 395.

45. See "Advertisements," *Straits Times Daily Advertiser*, 6 July 1900, 7; Makepeace, Brooke, and Braddell, *One Hundred Years*, 2:310.

46. "Recollections of Johore by T. P. Coe, November 1929 to May 1931," f. 3, RCS 103/4/3/2, CUL.

47. "Recollections of Thirty-Five Years as a Missionary Teacher in Malaya, by Reverend Preston Littlepage Peach, 1962," f. 8, RCS 103/4/26, CUL.

48. "Great Black Out in Singapore," *Singapore Free Press and Mercantile Advertiser*, 17 April 1936, 1.

49. Priya Jain, "Clean, Cool Air: Health and Air Conditioning in India (1920s–1960s)," *Journal of Architectural Education* 76:1 (2022): 134.

50. Raymond Arsenault, "The End of the Long Hot Summer: The Air Conditioner and Southern Culture," *Journal of Southern History* 50:4 (1984): 600–601.

51. Rudolf Mrázek, *Engineers of Happy Land: Technology and Nationalism in a Colony* (Princeton, NJ: Princeton University Press, 2002), 79.

52. "Modern Air-Conditioning System," *Morning Tribune*, 30 March 1936, 15.

53. "A Boon to the Worker," *Malaya Tribune*, 23 December 1937, 13.

54. R. E. Thompson, "Promoting Greater Efficiency and Less Fatigue," *Malaya Tribune*, 10 July 1939, 9.

55. "What Air-Conditioning Really Is," *Morning Tribune*, 15 June 1938, 6.

56. J. W. Gregory, "Inter-Racial Problems and White Colonisation in the Tropics," Report of the 92nd Meeting of the BAAS, 1924.

57. Jain, "Clean, Cool Air," 134.

58. Salvatore Basile, *Cool: How Air Conditioning Changed Everything* (New York: Fordham University Press, 2014), 3–4, 23–27.

59. "Air Conditioning," *Sunday Tribune*, 10 February 1935, 10.

60. Siry, *Air-Conditioning*, 12–18.

61. "Sea Breezes at the Capitol," *Straits Times* 29 April 1930, 19.

62. Locomotive Railway Supplement, *Moore's Monthly Magazine* 31 (1925): 110.

63. Sir Run Run Shaw, "Economic Development of Singapore," accession no. 001545, reel 1, Oral Histories, NAS.

64. Barbara Quek, "Cinema Pioneer Tan Cheng Kee," *BiblioAsia* 18:1 (2022): 54–57; "Attractive Lighting and a Modern Stage," *Morning Tribune*, 27 July 1938, 2.

65. "Weather Problems," *Morning Tribune*, 27 July 1938, 7.

66. "Grand New Alhambra Opens with Excellent Film," *Morning Tribune*, 28 July 1938, 3.

67. "Greetings from Java," *Morning Tribune*, 28 July 1938, 3.

68. Andrew Field, "Building Shanghai's Dreamworld: Architects and Elite Ballroom Designs of the 1920s and 1930s," *Built Heritage* 3 (2019): 34–48.

69. "Advert for the New World Dance Palace," *Morning Tribune*, 28 July 1938, 7. Wong Yunn Chii and Tan Kar Lin, "Emergence of a Cosmopolitan Space for Culture and Consumption: The New World Amusement Park–Singapore (1923–70) in the Inter-War Years," *Inter-Asia Cultural Studies* 5:2 (2004): 279–304.

70. James Nott, "Dance Halls: Towards and Architectural and Spatial History, c. 1918–65," *Architectural History* 61 (2018): 205–33.

71. "Singapore's Newest Cabaret," *Morning Tribune*, 16 May 1938, 2.

72. Wong Yunn Chii and Tan Kar Lin, "Emergence of a Cosmopolitan Space," 279.

73. "Advertisement," *Malaya Tribune*, 10 December 1937, 22.

74. "Air Conditioning," *Sunday Tribune*, 10 February 1935, 10.

75. "Singapore's Newest Cabaret," *Morning Tribune*, 16 May 1938, 2.

76. "System of Ventilation," *Malaya Tribune*, 2 August 1939, 5; Jiat Hwee Chang, "Before and behind the Pioneers of Modern Architecture in Singapore," *docomomo journal* 57 (2017): 60.

77. Bonny Tan, "Living It Up at the Capitol," *BiblioAsia* 13:4 (2018): 16–21.

78. "Air-Conditioned Capitol has 1,000 New Seats," *Straits Times*, 31 January 1940, 15.

79. Makepeace, Brooke, and Braddell, *One Hundred Years*, 2:199–200.

80. Anne Millbrooke, *Carrier in the Pacific and the Far East* (1983), historical report no. 32, box no. 154393, 3, United Technologies Corporation, Archive and Historical Resource Center, Carrier Corporation, Syracuse. The author wishes to thank Jiat Hwee Chang, National University of Singapore, for sharing this file.

81. Untitled article, *Malaya Tribune*, 24 June 1938, 8.

82. Millbrooke, *Carrier in the Pacific*, 34.

83. Goh Chor Boon, *Technology and Entrepôt Colonialism*, 171.

84. Peter Atkins, "School Milk in Britain, 1900–1934," *Journal of Policy History* 19:4 (2007): 395–427.

85. "Friesians Will Form Nucleus of Local Dairy Herd," *Straits Times*, 24 March 1946, 2; "Air Conditioning for Racehorses," *Malaya Tribune*, 20 December 1938, 2.

86. Kartar Singh, "The Public Service," accession no. 002335, reel/disc 13, transcript, p.163, Oral Histories, NAS.

87. Quek, "Tan Cheng Kee," 56.

88. Tan Cheng Hwee, "Japanese Occupation of Singapore," accession no. 000416, reels 7–8, transcript, pp. 79–80, 91–103, Oral Histories, NAS.

89. "An All-Electric World in 1963," *Sunday Tribune*, 15 May 1938, 18.

90. "Air-Conditioned Stand at Show," *Singapore Free Press and Mercantile Advertiser*, 8 August 1939, 3.

91. David Edgerton, *Shock of the Old: Technology and Global History since 1900* (London: Profile Books, 2008), ix–xviii.

92. "The War of the Future," *Straits Times*, 11 April 1930, 12.

93. "New Architecture in Singapore," *Straits Times*, 6 March 1938, 32.

94. Arsenault, "End of the Long Hot Summer," 600–601.

95. Jain, "Clean, Cool Air," 135. See also Mrázek, *Engineers of Happy Land*, 77–78.

96. Claire Lowrie, *Masters and Servants: Cultures of Empire in the Tropics* (Manchester: Manchester University Press, 2016).

97. Ronald Benjamin Milne, "Communities of Singapore (Part 1)," accession no. 000447, reel 1, transcript, pp. 1–3, and reel 3, p. 20, Oral Histories, NAS.

98. Niranjan Singh, "Communities of Singapore (Part 2)," accession no. 000440, reels 2, 71, Oral Histories, NAS. The new suburb is clearly marked on the "Survey Map of Singapore Town," media image no. 20120001645-0001, 1930, NAS.

99. Chan Kwee Sung, "Special Project," accession no. 000962, reel 1, p. 4; Mehervan Singh, "Communities of Singapore," accession no. 0005533, reel 6, both in Oral Histories, NAS.

100. "The Heat Wave," *Singapore Free Press*, 14 April 1926, 8.

101. "Heat Wave: No Break in Weather," *Malaya Tribune*, 11 June 1948, 1.

102. Kartar Singh, "The Public Service," accession no. 002335, reel/disc 13, transcript, pp. 156–64, Oral Histories, NAS; "Air-Conditioned Capitol Has 1,000 New Seats," *Straits Times*, 31 January 1940, 156–58.

103. Goh Chor Boon, *Technology and Entrepôt Colonialism*, 171–72.

104. Shennan, *Out in the Midday Sun*, 81–82.

105. Quote taken from Shennan, *Out in the Midday Sun*, 156.

106. Paul Carroll, "Special Project," accession no. 002273, reel/disc 4, transcript, p. 56, Oral Histories, NAS.

107. Chua Chye Chua, "Economic Development of Singapore," accession no. 002140, reel 6, audio file, 1 min. 24 sec., Oral Histories, NAS.

108. Dr. George Yuille Caldwell, "New Citizens," accession no. 003776, reel 12, audio file, 5 min. 15 sec., Oral Histories, NAS.

109. Teo Yew Chiat, "Chinese Dialect Groups," accession no. 003510, reel/disc 3, description, Oral Histories, NAS.

110. Shennan, *Out in the Midday Sun*, 157.

111. Goh Eng Wah, "Economic Development of Singapore," accession no. 001907, reel 5, Oral Histories, NAS.

112. Patrick James Kiely, "Education in Singapore (Part 1: English)," accession no. 003297, reel 1, Oral Histories, NAS.

113. David Penberthy, "Special Project," accession no. 003074, reel/disc 1, transcript, pp. 8–9, Oral Histories, NAS.

114. Millbrooke, *Carrier in the Pacific*, 36.

115. Millbrooke, *Carrier in the Pacific*, 44–46.

116. Chan Thye Seng, "The Public Service," accession no. 002265, reel 9, transcript, p. 105, Oral Histories, NAS.

117. Arnold, "Europe, Technology, and Colonialism," 96.

118. Frank Trippett, "The Great American Cooling Machine," *Time Magazine*, 13 August 1979, 75, quoted from Raymond Arsenault, "End of the Long Hot Summer," 597.

119. "Air Con," *Business Times*, 19 January 1999, 2; "Air-Con Gets My Vote, says SM Lee," *Straits Times*, 19 January 1999, 1; "Lee Kuan Yew: Strictly Singapore's Leadership Model," *Indian Economic Times*, 25 March 2015.

120. Eike-Christian Heine and Martin Meiske, "Scientific Bonanzas: Infrastructures as Places of Knowledge Production," in *Beyond the Lab and the Field: Scientific Infrastructures as Places of Knowledge Production since the Late Nineteenth Century*, ed. Eike-Christian Heine and Martin Meiske (Pittsburgh, PA: University of Pittsburgh Press, 2022), 3–20.

121. Smith, "Where There Is Life," 296. Smith quotes from Lim Thean Soo's *Ricky Star* (1978), 214.

122. Yuya Takane, Yukihiro Kikegawa, Masayuki Hara, and C. Sue B. Grimmond, "Urban Warming and Future Air-Conditioning Use in an Asian Megacity: Importance of Positive Feedback," *npj Climate and Atmospheric Science* 2:39 (2019), doi.org/10.1038/s41612-019-0096-2/.

CONCLUSION: THE DUSK OF EMPIRE

1. K. Rajendram, "Development Programme for Meteorological Services in Malaya and Singapore 1963–65," Report of the Conference of Commonwealth Meteorologists, Bracknell, 7–10 May 1963 (1964): 40, Bib. no. 571798, United Kingdom Meteorological Office Library; "Conference of Commonwealth Meteorologists," *Nature* 4883 (1963): 838–39.

2. Francis Kok Wah Loh, "Malaysia (The Federation of Malaysia): A Centralised Federation with Challenges from the States," in *The Forum of Federations Handbook of Federal Countries*, ed. A. Griffiths, R. Chattopadhyay, J. Light, and C. Stieren (London: Palgrave Macmillan, 2020), 201–13, esp. 203.

3. "Part II Posts," *Straits Times*, 28 November 1951, 4; *Research Facilities in Science and Technology in Asia: a Preliminary Survey* (UNESCO, 1968), 640–41.

4. "$2mil Changi Satellite Station Planned," *New Nation*, 25 January 1978, 4.

5. *WMO Bulletin* 17:2 (April 1968): 48; *WMO Bulletin* 19:1 (January 1970): 70.

6. Kanagasabai s/o Narayasamy (N. K. Sabai), "The Public Service," accession no. 001496, reel 1, transcript, pp. 4–5, Oral Histories, NAS.

7. *The Colonial Office List* (London: His Majesty's Stationary Office, 1951), 572.

8. "Local Man Is Weather Chief," *Malaya Tribune*, 22 July 1949, 2.

9. "Many Malayans Now Study in Australia," *Singapore Free Press*, 2 November 1954, 11.

10. "Meteorologist for Europe," *Straits Times*, 30 March 1963, 9; "Singapore Meteorologist at World Seminar in Australia," *Eastern Sun*, 24 December 1968, 3.

11. "Plea for an RAAF Plane," *Straits Times*, 19 June 1963, 6; "Rain-Making to Resume Today," *Straits Times*, 20 July 1963, 7.

12. "All-Day Wait in Vain from Suitable Clouds," *Straits Times*, 3 July 1963, 9; "Water Rationing to Continue," *Straits Budget*, 17 July 1963, 9.

13. "Science Was Ready to Make It Rain," *Straits Times*, 9 May 1963, 1.

14. "Rain-Making to Start Again," *Straits Times*, 23 August 1963, 4.

15. A. J. Morris, "Rocket Rain," *Straits Times*, 14 May 1963, 8.

16. "Report from the International Indian Ocean Expedition Conducted by Robert G. Snider, on Meteorological Facilities in the Eastern Half of the Indian

Ocean, 27 January to 3 March 1961," Scientific Committee on Oceanic Research, History Report 2 (2020): 3.

17. "Forecasting of Heavy Rains and Floods—Seminar in Kuala Lumpur, November 1968," *WMO Bulletin* 18:2 (1969): 114.

18. Hugh Mabbett, "Ending Monsoon Floods by Wringing Out the Clouds," *Straits Times*, 12 November 1968, 10. See also "Veteran Journalist Hugh Mabbett Dies of Cancer," *Straits Times*, 4 July 1990, 24.

19. Graham Sutton, "Steady Progress in Techniques of Forecasting," in *The World in 1984: The Complete New Scientist Series*, ed., Nigel Calder (Harmondsworth: Penguin Books, 1965), 1:115–19, esp. 116–19.

20. Yuriko Furuhata, *Transpacific Experiments in Atmospheric Control* (Durham, NC: Duke University Press, 2022), 57–58.

21. Mabbett, "Ending Monsoon Floods," 10.

22. "Latest Weather Forecast Gear for Our Airport," *Straits Times*, 3 December 1970, 2; "Regional Seminars and Technical Conferences," *WMO Bulletin* 20:1 (1971): 46; "Synoptic Analysis and Forecasting in the Tropics of Asia and the South-West Pacific," *WMO Bulletin* 20:2 (1971): 125–27.

23. "The World Weather Watch at 50," *WMO Bulletin* 62:1 (2013): 3–5.

24. GARP Special Report No. 28, Report of the Fourth Planning Meeting for the Monsoon Experiment (MONEX), Kuala Lumpur, 21–24 February 1978, Appendix E, Global Atmospheric Research Programme, WMO.

25. Films and Recordings from the UCAR Information Office, NCAR Archives, MONEX: the Monsoon Experiment, at https://opensky.ucar.edu/islandora/object/archives%3A9306/.

26. Richard S. Greenfield and T. N. Krishnamurti, "The Winter Monsoon Experiment—Report of the December 1978 Field Phase," *Bulletin of the American Meteorological Society* 60:5 (1979): 439–44, esp. 440.

27. Greenfield and Krishnamurti, "Winter Monsoon Experiment," 439–44.

28. "Flood Death Toll Raises to Seven," *Straits Times*, 6 December 1978, 1; "The Big Floods," four-page special pullout, *New Nation*, 4 December 1978, 1.

29. Loh Kah Seng and Michael D. Pante, "Controlling Nature, Disciplining Human Nature: Floods in Singapore and Metro Manila, 1945–1980s," *Nature and Culture* 10:1 (2015): 36–56.

30. Sanjay Subrahmanyam, "Connected Histories: Notes towards a Reconfiguration of Early Modern Eurasia," *Modern Asian Studies* 31:3 (1997): 735–62.

31. Jürgen Renn, *The Evolution of Knowledge: Rethinking Science for the Anthropocene* (Princeton, NJ: Princeton University Press, 2020), 15.

SELECTED BIBLIOGRAPHY

Achbari, Azadeh, and Frans van Lunteren. "Dutch Skies, Global Laws: The British Creation of Buys Ballot's Law." *Historical Studies in the Natural Sciences* 46:1 (2016): 1–43.

Ackernmann, Marsha E. *Cool Comfort: America's Romance with Air-Conditioning.* Washington, DC: Smithsonian Institution Press, 2002.

Adamson, George. "Imperial Oscillations: Gilbert Walker and the Construction of the Southern Oscillation." In *Weather, Climate, and the Geographical Imagination: Placing Atmospheric Knowledges,* edited by Martin Mahony and Sam Randalls, 43–66. Pittsburgh, PA: University of Pittsburgh Press, 2020.

Agaoglu, Erman, and Abu Hanifah Haris. "Technical Education in Malaya: The Development of Technical School Kuala Lumpur, 1925–1940." *Sejarah: Journal of History Department, University of Malaya* 29:2 (2020): 108–19.

Aiken, Robert. "Early Penang Hill Station." *Geographical Review* 77:4 (1987): 422–25.

Allan, Robert J., and Rosanne D. D'Arrigo. "'Persistent' ENSO Sequences: How Unusual Was the 1990–1995 El Niño?" *The Holocene* 9:1 (1999): 101–18.

Allan, Robert J., Joëlle Gergis, and Rosanne D'Arrigo. "Placing the AD 2014–2016 'Protracted' El Niño into a Long-Term Context." *The Holocene* 30:1 (2020): 90–105.

Allen, Michael Thad, and Gabrielle Hecht. "Authority, Political Machines, and Technology's History." In *Technologies of Power: Essays in Honour of Thomas Parke Hughes and Agatha Chipley Hughes,* edited by Michael Thad Allen and Gabrielle Hecht, 1–24. Cambridge, MA: MIT Press, 2001.

Alphonso, G., and Albert Lau. *The Causeway.* Kuala Lumpur and Singapore: National Archives of Malaysia and Singapore, 2011.

Alvarez, Kerby. "Instrumentation and Institutionalization: Colonial Science and the Observatorio Meteorológico de Manila, 1865–1899." *Philippine Studies: Historical and Ethnographic Viewpoints* 63:3–4 (2016): 385–416.

Amrith, Sunil. *Unruly Waters: How Rains, Coasts, and Seas Have Shaped Asia's History*. London: Penguin, 2018.

Anderson, Katherine. "Looking at the Sky: The Visual Context of Victorian Meteorology." *British Journal for the History of Science* 36:3 (2003): 301–32.

Anderson, Katherine. *Predicting the Weather: Victorians and the Science of Meteorology*. Chicago: Chicago University Press, 2005.

Anderson, Warwick. *Colonial Pathologies: American Tropical Medicine, Race, and Hygiene in the Philippines*. Durham, NC: Duke University Press, 2006.

Anduaga, Aitor. *Cyclones and Earthquakes: The Jesuits, Prediction, Trade, and Spanish Dominion in Cuba and the Philippines, 1850–1898*. Manila: Ateneo de Manila University Press, 2017.

Anduaga, Aitor. *Politics, Statistics and Weather Forecasting, 1840–1910: Taming the Weather*. London: Routledge, 2020.

Anduaga, Aitor. "Transnational Co-production of Knowledge: The Standardisation of Typhoon Warning Codes in the Far East, 1900–1939." *Minerva* (2022): 1–23.

Arnold, David. *Colonizing the Body: State Medicine and Epidemic Disease in Nineteenth Century India*. Berkeley: University of California Press, 1993.

Arnold, David. "Europe, Technology, and Colonialism in the 20th Century." *History and Technology* 21:1 (2005): 85–106.

Arnold, David. *The Tropics and the Traveling Gaze: India, Landscape and Science, 1800–1856*. Seattle: University of Washington Press, 2006.

Arnold, David, and Ramachandra Guha, eds. *Nature, Culture, Imperialism: Essays on the Environmental History of South Asia*. Oxford: Oxford University Press, 1996.

Arsenault, Raymond. "The End of the Long Hot Summer: The Air Conditioner and Southern Culture." *Journal of Southern History* 50:4 (1984): 597–628.

Aubin, David, Charlotte Bigg, and Otto H. Sibum, eds. *Observatories and Astronomy in Nineteenth-Century Science and Culture*. Durham, NC: Duke University Press, 2010.

Bahauddin, Azizi, and Teh Boon Soon. "The Environment and *Feng Shui* Application in Cheong Fatt Tze Mansion, Penang, Malaysia." In *WIT Transactions on The Built Environment*, 183:1–13. Ashurst: WIT Press, 2019.

Baldwin, Peter. *Contagion and the State in Europe, 1830–1930*. Cambridge: Cambridge University Press, 2004.

Ballantyne, Andrew, and Andrew Law. "Genealogy of the Singaporean Black-and-White House." *Singapore Journal of Tropical Geography* 32 (2011): 301–13.

Bankoff, Greg. "Aeolian Empires: The Influence of Winds and Currents on European Maritime Expansion in the Days of Sail." *Environment and History* 23:2 (2017): 163–96.

Bankoff, Greg, and Peter Boomgaard, eds. *A History of Natural Resources in Asia: The Wealth of Nature*. London: Palgrave Macmillan, 2007.

Barber, Daniel A. *Modern Architecture and Climate: Design before Air-Conditioning*. Princeton, NJ: Princeton University Press, 2020.

Barnard, Timothy P. "Celates, Raya-Laut, Pirates: The Orang Laut and Their Decline in History." *JMBRAS* 80:2 (2007): 33–49.

Barnard, Timothy P., ed. *Nature Contained: Environmental Histories of Singapore*. Singapore: NUS Press, 2014.

Barnard, Timothy P. *Nature's Colony: Empire, Nation and Environment in the Singapore Botanic Gardens*. Singapore: NUS Press, 2016.

Barnard, Timothy P., and Joanna W. C. Lee. "A Spiteful Campaign: Agriculture, Forests, and Administering the Environment in Imperial Singapore." *Environmental History* 27:3 (2022): 467–90.

Barton, Gregory A. *Empire Forestry and the Origins of Environmentalism*. Cambridge: Cambridge University Press, 2002.

Barton, Gregory A. "Empire Forestry and the Origins of Environmentalism." *Journal of Historical Geography* 27:4 (2001): 529–52.

Basile, Salvatore. *Cool: How Air Conditioning Changed Everything*. New York: Fordham University Press, 2014.

Bassett, D. K. "British Trade and Policy in Indonesia, 1760–1772." *Bijdragen tot de Taal, Land en Volkenkunde* 120, 2de Afl. (1964): 197–223.

Bayley, Christopher, and Tim Harper. *Forgotten Armies: The Fall of British Asia, 1941–1945*. Cambridge, MA: Harvard University Press, 2005.

Beattie, James. *Empire and Environmental Anxiety: Health, Science, Art and Conservation in South Asia and Australasia, 1800–1920*. Basingstoke: Palgrave Macmillan, 2011.

Beattie, James, and Ruth Morgan. "Engineering Edens on This 'Rivered Earth'? A Review Article on Water Management and Hydro-Resilience in the British Empire, 1860s–1940s." *Environment and History* 23 (2017): 39–63.

Beattie, James, Emily O'Gorman, and Matthew Henry, eds. *Climate, Science, and Colonization: Histories from Australia and New Zealand*. New York: Palgrave Macmillan, 2014.

Beinart, William, and Lotte Hughes. *Environment and Empire*. Oxford: Oxford University Press, 2007.

Bell, Morag. *Geography and Imperialism, 1820–1940*. Manchester: Manchester University Press, 1995.

Bennett, Brett M. *Plantations and Protected Areas: A Global History of Forest Management*. Cambridge, MA: MIT Press, 2015.

Bennett, Brett M., and Joseph M. Hodge. *Science and Empire: Knowledge and Networks of Science across the British Empire, 1800–1970*. Basingstoke: Palgrave Macmillan, 2011.

Bhattacharya, Nandini. *Contagion and Enclaves: Tropical Medicine in Colonial India*. Liverpool: Liverpool University Press, 2012.

Blackburn, Kevin, and Karl Hack. *Did Singapore Have to Fall? Churchill and the Impregnable Fortress*. London: Routledge, 2004.

Bonney, R. "Francis Light and Penang." *JMBRAS* 38:1 (1965): 135–58.

Boomgaard, Peter, ed. *Empire and Science in the Making: Dutch Colonial Scholarship in Comparative Perspective, 1760–1830*. New York: Palgrave Macmillan, 2013.

Braddell, Roland. *The Lights of Singapore*. 2nd ed. London: Methuen, 1934. Reprint, Oxford: Oxford University Press, 1982.

Broich, John. "Engineering the Empire: British Water Supply Systems and Colonial Societies, 1850–1900." *Journal of British Studies* 46:2 (2007): 346–65.

Buckley, Charles Burton. *An Anecdotal History of Old Times in Singapore*. 1902. 2 vols. Reprint, Singapore: Forgotten Books, 1984.

Burton, Jim. "Robert Fitzroy and the Early History of the Meteorological Office." *British Journal for the History of Science* 19:2 (1986): 147–76.

Calder, Nigel. *The World in 1984: The Complete New Scientist Series*. Vol. 1. Harmondsworth: Penguin Books, 1965.

Carson, Sarah. "Anticipating the Monsoon: The Necessity and Impossibility of the Seasonal Weather Forecast for South Asia, 1886–1953." *British Journal for the History of Science* 54:3 (2021): 305–25.

Carter, C. R. *Magnetic Fever: Global Imperialism and Empiricism in the Nineteenth Century*. Philadelphia, PA: American Philosophical Society, 2009.

Cawood, J. "The Magnetic Crusade: Science and Politics in Early Victorian Britain." *Isis* 70:4 (1979): 492–518.

Chang, Hasok. *Inventing Temperature: Measurement and Scientific Progress*. Oxford: Oxford University Press, 2004.

Chang, Jiat Hwee. "Before and behind the Pioneers of Modern Architecture in Singapore." *docomomo journal* 57 (2017): 57–63

Chang, Jiat Hwee, and Anthony D. King. "Towards a Genealogy of Tropical Architecture: Historical Fragments of Power-Knowledge, Built Environment and Climate in the British Colonial Territories." *Singapore Journal of Tropical Geography* 32 (2011): 283–300.

Chenoweth, Mike, J. M. Vaquero, Ricardo Garcia-Herrera, and Dennis Wheeler. "A Pioneer in Tropical Meteorology: William Sharpe's Barbados Weather Journal, April–August 1680." *Bulletin of the American Meteorological Society* 88:12 (2007): 1957–64.

Chinnici, Ileana, and Guy Consolmagno, eds. *Angelo Secchi and Nineteenth Cen-*

tury Science: The Multidisciplinary Contributions of a Pioneer and Innovator. Cham: Springer Nature Switzerland, 2021.

Coen, Deborah. *Climate in Motion: Science, Empire, and the Problem of Scale.* Chicago: University of Chicago Press, 2018.

Cohn, Bernard. *Colonialism and Its Forms of Knowledge: The British in India.* Princeton, NJ: Princeton University Press, 1996.

Comber, Leon. "Chinese Secret Societies in Malaya: An Introduction." *Journal of the Malayan Branch of the Royal Asiatic Society* 29:1 (1956): 146–62.

Cook, E. R., K. J. Anchukaitis, B. M. Buckley, R. D. D'Arrigo, G. C. Jacoby, and W. E. Wright. "Asian Monsoon Failure and Megadrought during the Last Millennium." *Science* 328 (2010): 486–89.

Corlett, Richard T. "The Ecological Transformation of Singapore, 1819–1990." *Journal of Biogeography* 19:4 (1992): 411–20.

Corner, E. J. H. *The Marquis: A Tale of Syonan-to.* Singapore: Heinemann Asia, 1981.

Corner, John K. *My Father in His Suitcase: In Search of E. J. H. Corner, the Relentless Botanist.* Singapore: Landmark Books, 2013.

Courtney, Chris. "The Birth of Cool: Heat and Air-Conditioning in the History of Wuhan, 1950–2020." *Urban Studies* (2023): 1–16. doi.org/10.1177/00420980231153512/.

Crawfurd, J. *Journal of an Embassy to the Courts of Siam and Cochin-China.* 2 vols. London, 1828.

Crowley, John E. *The Invention of Comfort: Sensibilities and Design in Early Modern Britain and Early America.* Baltimore: Johns Hopkins University Press, 2001.

Cushman, Gregory T. "Humboldtian Science, Creole Meteorology, and the Discovery of Human-Caused Climate Change in South America." *Osiris* 26 (2011): 19–44.

Dale, O. J. *Urban Planning in Singapore: Transformation of a City.* Malaysia: Oxford University Press, 1999.

Daniell, John F. *Elements of Meteorology.* 3rd ed., vol. 1. London: John W. Parker, 1845.

Dassow Walls, Laura. *The Passage to Cosmos: Alexander von Humboldt and the Shaping of America.* Chicago: University of Chicago Press, 2009.

Davis, Mike. *Late Victorian Holocausts: El Niño Famines and the Making of the Third World.* London: Verso, 2001.

Deb Roy, Rohan. *Malarial Subjects: Empire, Medicine, and Nonhumans in British India, 1820–1909.* Cambridge: Cambridge University Press, 2017.

Dobbs, Stephen. *The Singapore River: A Social History, 1819–2002.* Singapore: NUS Press, 2003.

Drabble, J. H., "The Plantation Rubber Industry in Malaya up to 1922." *JMBRAS* 40:1 (1967): 52–77.

Driver, Felix, and Luciana Martins, eds. *Tropical Visions in an Age of Empire*. Chicago: Chicago University Press, 2005.

D'Souza, Rohan. *Drowned and Damned: Colonial Capitalism and Flood Control in Eastern India*. Delhi: Oxford University Press India, 2016.

D'Souza, Rohan. "Event, Process and Pulse: Resituating Floods in Environmental Histories of South Asia." *Environment and History* 26:1 (2020): 31–50.

Edgerton, David. *Shock of the Old: Technology and Global History since 1900*. London: Profile Books, 2008.

Edwards, Paul N. "Meteorology as Infrastructural Globalism." *OSIRIS* 21 (2006): 229–50.

Edwards, Paul N. *A Vast Machine: Computer Models, Climate Data, and the Politics of Global Warming*. Cambridge, MA: MIT Press, 2010.

Elvin, Mark. *The Retreat of the Elephants: An Environmental History of China*. New Haven, CT: Yale University Press, 2004.

Endfield, Georgina, and Carol Morris. "Cultural Spaces of Climate." *Climatic Change* 113 (2012): 1–4.

Endo, Nobuhiko, and Jun Matsumoto. "A Brief History of Meteorological Observations in French Indochina." *Geographical Reports of Tokyo Metropolitan* 54 (2019): 33–42.

Evans, Geoffrey. "The Rubber Research Institute of Malaya." *Nature* 3714 (1941): 15–16.

Fichman, Martin. *An Elusive Victorian: The Evolution of Alfred Russel Wallace*. Chicago: Chicago University Press, 2004.

Field, Andrew. "Building Shanghai's Dreamworld: Architects and Elite Ballroom Designs of the 1920s and 1930s." *Built Heritage* 3 (2019): 34–48.

Fielding, K. J. "The Settlement of Penang by James Scott." *JMBRAS* 28:1 (1955): 37–51.

Fleming, James R. *The Callendar Effect: The Life and Work of Guy Stewart Callendar, 1898–1964*. Boston, MA: American Meteorological Association, 2007.

Fleming, James R. "The Climate Engineers." *Wilson Quarterly* 31:2 (2007): 46–60.

Fleming, James R. *First Woman: Joanne Simpson and the Tropical Atmosphere*. Oxford: Oxford University Press, 2020.

Fleming, James R. *Fixing the Sky: The Chequered History of Weather and Climate Control*. New York: Columbia University Press, 2010.

Fleming, James R. *Inventing Atmospheric Science: Bjerknes, Rossby, Wexler, and the Foundations of Modern Meteorology*. Cambridge, MA: MIT Press, 2016.

Fleming, James R., and Vladimir Jankovic. "Introduction: Revisiting Klima." *Osiris* 26 (2011): 1–15.

Fleming, James R., Vladimir Jankovic, and Deborah R. Coen, ed. *Intimate Universality: Local and Global Themes in the History of Weather and Climate*. Sagamore Beach, MA: Science History Publications, 2006.

Frank, Mark E. "Frontier Atmosphere: Observation and Regret at Chinese Weather Stations in Tibet, 1939–1949." *British Journal for the History of Science* 54:3 (2021): 361–79.

Friedman, Robert Marc. *Appropriating the Weather: Vilhelm Bjerknes and the Construction of a Modern Meteorology*. Ithaca: Cornell University Press, 1989.

Furuhata, Yuriko, *Transpacific Experiments in Atmospheric Control*. Durham, NC: Duke University Press, 2022.

Gao, Elaine, Bertrand Timbal, and Fiona Williamson. "Creating Singapore's Longest Monthly Rainfall Record from 1839 to the Present." *MSS Research Letters* 2 (2018): 3–11.

Garcia-Herrera, Ricardo, C. Wilkinson, F. B. Koek, M. R. Prieto, N. Calvo, and E. Hernández. "Description and General Background to Ships' Logbooks as a Source of Climatic Data." *Climatic Change* 73:1 (2005): 13–36.

Garden, Don. *Droughts, Floods & Cyclones: El Niños That Shaped Our Colonial Past*. Melbourne: Australian Scholarly Publishing, 2009.

George, Cherian. *Singapore: The Air-Conditioned Nation: Essays on the Politics of Comfort and Control, 1990–2000*. Singapore: Landmark Books, 2000.

Gergis, Joëlle L., Karl Braganza, Anthony Fowler, Scott Mooney, and James Risbey. "Reconstructing El Niño–Southern Oscillation (ENSO) from High-Resolution Palaeoarchives." *Journal of Quarternary Science* 21:7 (2006): 707–22.

Gergis, Joëlle L., and A. M. Fowler. "A History of ENSO Events since AD 1525: Implications for Future Climate Change." *Climatic Change* 92 (2009): 343–87.

Ghosh, Amitav. *The Nutmeg's Curse: Parables for a Planet in Crisis*. Chicago: University of Chicago Press, 2021.

Gibson-Hill, C. A. "Singapore Old Strait and New Harbour, 1300–1870." *Memoirs of the Raffles Museum* (Singapore: Government Printing Office, 1956), 3:1–115.

Gilmartin, David. "Scientific Empire and Imperial Science: Colonialism and Irrigation Technology in the Indus Basin." *Journal of Asian Studies* 53:4 (1994): 1127–49.

Gilmour, Oswald. *With Freedom to Singapore: A Social and Administrative Assessment of Post-War Singapore, 1945–1950*. 1950. Reprint, London: Ernest Benn, 2020.

Glacken, C. J. *Traces on the Rhodian Shore: Nature and Culture in Western Thought from Ancient Times to the End of the Eighteenth Century*. Berkeley: University of California Press, 1967.

Goh Chor Boon. "Imported Technology: Its Idea and Development." *JMBRAS* 71:1 (1998): 41–54.

Goh Chor Boon. *Technology and Entrepôt Colonialism in Singapore, 1819–1940*. Singapore: ISEAS Publishing, 2013.

Graber, Frédéric. "Inventing Needs: Expertise and Water Supply in Late Eighteenth- and Early Nineteenth-Century Paris." *British Journal for the History of Science* 40:3 (2007): 315–32.

Goldingham, J. *Madras Observatory Papers*. Madras: College Press of Fort St. George, 1827.

Golinski, Jan. "Barometers of Change: Meteorological Instruments as Machines of Enlightenment." In *The Sciences in Enlightened Europe*, edited by William Clark, Jan Golinski, and Simon Schaffer, 50–69. Chicago: University of Chicago Press, 1999.

Golinski, Jan. *Making Natural Knowledge: Constructivism and the History of Science*. Cambridge: Cambridge University Press, 1998.

Goodman, Matthew. "Follow the Data: Administering Science at Edward Sabine's Magnetic Department, Woolwich, 1841–57." *Notes and Records: The Royal Society Journal of the History of Science* 73:2 (2019): 187–202.

Goss, Andrew. *The Floracrats: State-Sponsored Science and the Failure of the Enlightenment in Indonesia*. Madison: University of Wisconsin Press, 2011.

Goss, Andrew. "Introduction: An Imperial Turn in the History of Science." In *The Routledge Handbook of Science and Empire*, edited by Andrew Goss, 1–9. London: Routledge, 2021.

Grove, Richard. *Ecology, Climate, and Empire: Colonialism and Global Environmental History, 1400–1940*. Cambridge: Cambridge University Press, 1997.

Grove, Richard. *Green Imperialism: Colonialism, Tropical Island Edens and the Origins of Environmentalism, 1600–1860*. Cambridge: Cambridge University Press, 1995.

Grove, Richard, and George Adamson. *El Niño in World History*. Basingstoke: Palgrave, 2017.

Grove, Richard, and J. Chappell. *El Niño: History and Crisis: Studies from the Asia-Pacific Region*. Cambridge, UK: White Horse Press, 2000.

Guha, Ramachandra. *The Unquiet Woods: Ecological Change and Peasant Resistance in the Himalaya*. Berkeley: University of California Press, 2000.

Gullick, J. M. "A Short History of the Society." *JMBRAS* 68:2 (1995): 67–79.

Gullick, J. M. "William Maxwell and the Study of Malay Society." *JMBRAS* 64:2 (1991): 7–46.

Halliday, Stephen. "Death and Miasma in Victorian London: An Obstinate Belief." *British Medical Journal* 323:7327 (2001):1469–71.

Hamblyn, Richard. *The Invention of Clouds: How an Amateur Meteorologist Forged the Language of the Skies*. London: Picador, 2011.

Hanitsch, R. N. Wallich, James Lumsdaine, and Francis Salmond. "Letters of Nathaniel Wallich relating to the Establishment of Botanical Gardens in Singapore." *JMBRAS* 42:1 (1969): 145–54.

Harley, J. B., and David Woodward, eds. *The History of Cartography*. Chicago: University of Chicago Press, 1987.

Harper, Kristine C. *Make It Rain: State Control of the Atmosphere in Twentieth-Century America*. Chicago: University of Chicago Press, 2017.

Harper, Kristine C. *Weather by the Numbers: The Genesis of Modern Meteorology*. Cambridge, MA: MIT Press, 2012.

Harrison, Mark. *Climates and Constitutions: Health, Race, Environment and British Imperialism in India, 1600–1850*. New Delhi: Oxford University Press, 1999.

Harrison, Mark. *Public Health in British India: Anglo-Indian Preventative Medicine 1859–1914*. Cambridge: Cambridge University Press, 1994.

Heffernan, M. "Bringing the Desert to Bloom: French Ambitions in the Sahara during the Late Nineteenth Century—The Strange Case of la mer intérieure." In *Water, Engineering and Landscape: Water Control and Landscape Transformation in the Modern Period*, edited by D. E. Cosgrove and G. E. Petts, 94–114. London: Belhaven Press 1990.

Heine, Eike-Christian, and Martin Meiske. "Scientific Bonanzas: Infrastructures as Places of Knowledge Production." In *Beyond the Lab and the Field: Scientific Infrastructures as Places of Knowledge Production since the Late Nineteenth Century*, edited by Eike-Christian Heine and Martin Meiske, 3–20. Pittsburgh, PA: University of Pittsburgh Press, 2022.

Hendel, Martin, Sophie Parison, Arnaud Grados, and Laurent Royon. "Which Pavement Structures Are Best Suited to Limiting the UHI Effect? A Laboratory-Scale Study of Parisian Pavement Structures." *Building and Environment* 144:15 (2018): 216–29.

Henry, Matthew. "Assembling the Weather: Expertise, Authority and the Negotiation of Trans-Tasman Aviation Forecasts." *History of Meteorology* 8 (2017): 179–201.

Hewitt, K., ed. *Interpretations of Calamity: From the Viewpoint of Human Ecology*. London: Routledge, 2021.

Higgitt, Rebekah. "A British National Observatory: The Building of the New Physical Observatory at Greenwich, 1889–1898." *British Journal for the History of Science* 47:4 (2013): 609–35.

Hill, R. D. *Rice in Malaya: A Study in Historical Geography*. Singapore: NUS Press, 2012.

Hills, Jill. *The Struggle for Control of Global Communication: The Formative Century*. Urbana: University of Illinois Press, 2010.

Hochadel, Oliver, and Agusti Nieto-Galen, eds. *Urban Histories of Science: Making Knowledge in the City, 1820–1940*. London: Routledge, 2018.

Hodge, Joseph M. *Triumph of the Expert: Agrarian Doctrines of Development and the Legacies of British Colonialism*. Athens: Ohio University Press, 2007.

Hollen Lees, Lynn. *Planting Empire, Cultivating Subjects: British Malaya, 1786–1941*. Cambridge: Cambridge University Press, 2017.

Holmwood, John, and John Scott, eds. *The Palgrave Handbook of Sociology in Britain*. Basingstoke: Palgrave Macmillan, 2014.

Huang, Chi Chi. "'Hong Kong Can Afford a Typhoon or Two': British Discussions of Revolving Storms." *British Journal for the History of Science* (2021): 1–13.

Hulme, Mike. "Climate and Its Changes: A Cultural Appraisal." *Geography and Environment* 2:1 (2015): 1–11.

Hussin, Nordin. *Trade and Society in the Straits of Melaka: Dutch Melaka and English Penang, 1780–1830*. Singapore: NUS Press, 2006.

Jain, Priya. "Clean, Cool Air: Health and Air Conditioning in India (1920s–1960s)." *Journal of Architectural Education* 76:1 (2022): 133–37.

Jarman, Robert L., ed. *Annual Reports of the Straits Settlements, 1855–1941*. 12 vols. Slough: Archive Editions, 1998.

Jasanoff, Sheila. "Future Imperfect: Science, Technology, and the Imaginations of Modernity." In *Sociotechnical Imaginaries and the Fabrication of Power*, edited by Sheila Jasanoff and Sang-Hyun Kim, 1–33. Chicago: University of Chicago Press, 2015.

Jenkins, Gwynn. *Contested Space: Cultural Heritage and Identity Reconstructions: Conservations Strategies within a Developing Asian City*. Munster: LIT Verlag, 2008.

Jeon, Chihyung. "Flying Weather Men and Robot Observers: Instruments, Inscriptions, and Identities in US Upper-Air Observation, 1920–1940." *History and Technology* 26:2 (2010): 119–45.

Jungnickel, Christa, and Russell McCormmach. *Intellectual Mastery of Nature: Theoretical Physics from Ohm to Einstein. The Torch of Mathematics, 1800–1870*. Vol. 1. Chicago: Chicago University Press, 1990.

Karakiewicz, J., and T. Kvan. *The Making of Hong Kong: From Vertical to Volumetric*. London: Routledge, 2010.

Karim, Wazir Jahan. "The 'Discovery' of Penang Island at Tanjong Tokong before 1785: Bapu Alaidin Meera Hussein Lebai and Captain Francis Light." *JMBRAS* 86:1 (2013): 1–29.

Kathirithamby-Wells, Jeyamalar. "The Implications of Plantation Agriculture for Biodiversity in Peninsular Malaysia: A Historical Analysis." In *Beyond the Sacred Forest: Complicating Conservation in Southeast Asia*, edited by Michael R. Dove, Percy E. Sajise, and Amity A. Doolittle, 62–90. Durham: Duke University Press, 2011.

Kathirithamby-Wells, Jeyamalar. *Nature and Nation: Forests and Development in Peninsular Malaysia*. Singapore: Nordic Institute of Asian Studies and NUS Press, 2005.

Kennedy, Dane. "The Perils of the Midday Sun: Climatic Anxieties in the Colonial Tropics." In *Imperialism and the Natural World*, edited by J. M. MacKenzie, 118–40. Manchester: Manchester University Press, 1990.

Kent, Randolph, and John Ratcliffe. *Responding to Catastrophes: U.S. Innovation in a Vulnerable World*. Washington, DC: CSIS Press, 2008.

Khoo Kay Kim. "Developments Relevant to Malayan Agriculture in the Post-rubber Crisis Era (1920–1921)." *JMBRAS* 72:2 (1999): 17–47.

Khoo Salma Nasution. *The Chulia in Penang: Patronage and Place-Making around the Kapitan Kling Mosque, 1786–1957.* Penang: Areca Books, 2014.

Kierner, Cynthia A. *Inventing Disaster: The Culture of Calamity from the Jamestown Colony to the Johnstown Flood.* Chapel Hill: University of North Carolina Press, 2019.

King, A. W. "Plantation and Agriculture in Malaya, with Notes on the Trade of Singapore." *Geographical Journal* 93:2 (1939): 136–48.

Kishore, Raghav. "'Urban 'Failures': Municipal Governance, Planning and Power in Colonial Delhi, 1863–1910." *Indian Economic and Social History Review* 52:4 (2015): 439–61.

Knobloch, Eberhard. "Alexander von Humboldt—The Explorer and the Scientist." *Centaurus* 49 (2007): 3–14.

Kohler, Robert E. *Landscapes and Labscapes: Exploring the Lab–Field Border in Biology.* Chicago: Chicago University Press, 2002.

Kohn, Edward P. *Hot Time in the Old Town: The Great Heat Wave of 1896 and the Making of Theodore Roosevelt.* New York: Basic Books, 2010.

Koike, Kenji. "Managing Agency Capitalism and Malayan Rubber: Harrisons & Crosfield, Ltd. (1900–1940)." *JMBRAS* 90:312 (2017): 73–100.

Kratoska, Paul H. "Imperial Unity versus Local Autonomy: British Malaya and the Depression of the 1930s." In *Weathering the Storm: The Economies of Southeast Asia in the 1930s Depression,* edited by Peter Boomgaard and Ian Brown, 271–94. Singapore: Institute of Southeast Asian Studies, 2000.

Kratoska, Paul H. *The Japanese Occupation of Malaya and Singapore, 1941–45: A Social and Economic History.* Singapore: NUS Press, 2018.

Krieger, Martin. *Nathaniel Wallich: Global Botany in Nineteenth-Century India.* Abingdon: Routledge, 2023.

Kumar, Deepak, Vinita Damodaran, and Rohan D'Souza, ed. *The British Empire and the Natural World: Environmental Encounters in South Asia.* New Delhi: Oxford University Press, 2011.

Kutzbach, G. *The Thermal Theory of Cyclones: A History of Meteorological Thought in the Nineteenth Century.* Boston, MA: American Meteorological Society, 1979.

Lambert, David, and Alan Lester, eds. *Colonial Lives across the British Empire: Imperial Careering in the Long Nineteenth Century.* Cambridge: Cambridge University Press, 2006.

Langdon, Marcus. *Penang: The Fourth Presidency of India, 1805–1830, Volume Two: Fire, Spice, and Edifice.* Penang: George Town World Heritage, 2015.

Lee, Kah-Wee, and Lawrence Chua. "Modernity's Other." *South East Asia Research* 28:2 (2020): 111–19.

Legg, Stephen. "Debating the Climatological Role of Forests in Australia, 1827–

1949: A Survey of the Popular Press." In *Climate, Science, and Colonization: Histories from Australia and New Zealand*, edited by James Beattie, Emily O'Gorman, and Matthew Henry, 119–36. New York: Palgrave Macmillan, 2014.

Lehmann, Philipp. *Desert Edens: Colonial Climate Engineering in an Age of Anxiety*. Princeton, NJ: Princeton University Press, 2022.

Lester, Alan. "Imperial Circuits and Networks: Geographies of the British Empire." *History Compass* 4 (2006): 124–41.

Lieberman, Victor, and Brendan Buckley. "The Impact of Climate on Southeast Asia, c. 950–1820: New Findings." *Modern Asian Studies* 46:5 (2012): 1049–96.

Little, Robert. "The Medical Topography of Singapore." *Journal of the Indian Archipelago and Eastern Asia* 3:8 (1848): 449–97.

Little, Robert. "Part I: On the medical Topography of Singapore, particularly its Marshes and Malaria." *Journal of the Indian Archipelago and Eastern Asia* 5 (1851): 449–71.

Liu, Gretchen. *Singapore: A Pictorial History, 1819–2000*. London: Psychology Press, 2001.

Livingstone, David N. *Putting Science in Its Place: Geographies of Scientific Knowledge*. Chicago: Chicago University Press, 2004.

Livingstone, David N. "Tropical Climate and Moral Hygiene: The Anatomy of a Victorian Debate." *British Journal for the History of Science* 32:1 (1999): 93–110.

Locher, Fabien, and Jean-Baptiste Fressoz. "Modernity's Frail Climate: A Climate History of Environmental Reflexivity." *Critical Inquiry* 38 (2012): 579–98.

Loh Kah Seng and Li Yan Hsu. *Pandemics in Singapore, 1819–2022: Lessons for the Age of Covid-19*. Abingdon: Routledge, 2024.

Loh Kah Seng and Michael D. Pante. "Controlling Nature, Disciplining Human Nature: Floods in Singapore and Metro Manila, 1945–1980s." *Nature and Culture* 10:1 (2015): 36–56.

Louis, William Roger. *British Strategy in the Far East, 1919–1939*. Oxford: Clarendon Press, 1971.

Lowenthal, David, ed. *George Perkins Marsh, Man and Nature: Or, Physical Geography as Modified by Human Action, 1864*. Seattle: University of Washington Press, 2003.

Lowrie, Claire. *Masters and Servants: Cultures of Empire in the Tropics*. Manchester: Manchester University Press, 2016.

Lowrie, Claire, Nicole Tarulevicz, Fiona Williamson, and Charmaine Lam, "Cosmopolitan Elites and the Management of Tropical Heat in Colonial Malaya, c. 1870s–1930s." *Journal of Colonial History and Colonialism* 25:1 (2024). doi.org/10.1353/cch.2024.a928069/.

Macdonald, Lee T. *Kew Observatory and the Evolution of Victorian Science, 1840–1910*. Pittsburgh, PA: University of Pittsburgh Press, 2018.

MacKeown, Kevin. *Early China Coast Meteorology: The Role of Hong Kong*. Hong Kong: Hong Kong University Press, 2011.

Macleod, Roy. "Passages in Imperial Science: From Empire to Commonwealth." *Journal of World History* 4:1 (1993): 117–50.

Mahony, Martin. "For an Empire of 'All Types of Climate': Meteorology as an Imperial Science." *Journal of Historical Geography* 51 (2016): 29–39.

Mahony, Martin. "The 'Genie of the Storm': Cyclonic Reasoning and the Spaces of Weather Observation in the Southern Indian Ocean, 1851–1925." *British Journal for the History of Science* 51:4 (2018): 607–33.

Mahony, Martin. "Historical Geographies of the Future: Airships and the Making of Imperial Atmospheres." *Annals of the American Association of Geographers* 109:4 (2019): 1279–99.

Mahony, Martin, and Georgina Endfield. "Climate and Colonialism." *WIREs Climate Change* 2018. 9:e510.

Mahony, Martin, and Samuel Randalls, eds. *Weather, Climate, and the Geographical Imagination: Placing Atmospheric Knowledges*. Pittsburgh, PA: University of Pittsburgh Press, 2020.

Makepeace, Walter, Gilbert E. Brooke, and Roland St. John Braddell, eds. *One Hundred Years of Singapore*. London: John Murray, 1921. 2 vols.

Manderson, Lenore. "Health Services and the Legitimation of the Colonial State: British Malaya, 1786–1941." *International Journal of Health Services* 17:1 (1987): 91–112.

Martin, D. C. "The Gassiot Committee of the Royal Society and Meteorological Research." *Nature* 190:4772 (1961): 212–13.

McAleer, John. "Stargazers at the World's End: Telescopes, Observatories, and 'Views' of Empire in the Nineteenth-Century British Empire." *British Journal for the History of Science* 46:3 (2011): 389–413.

McCrum, Ronald. *Men Who Lost Singapore, 1938–1942*. Singapore: NUS Press, 2017.

McCluskey, Michael, and Luke Seaber, eds. *Aviation in the Literature and Culture of Interwar Britain*. Cham, Switzerland: Palgrave MacMillan, 2021.

McCormmach, Russell. *Speculative Truth: Henry Cavendish, Natural Philosophy, and the Rise of Modern Theoretical Science*. Oxford: Oxford University Press, 2004.

McKittrick, Meredith. "An Empire of Rivers: The Scheme to Flood the Kalahari, 1919–1945." *Journal of Southern African Studies* 41:3 (2015): 485–504.

McKittrick, Meredith. "Talking about the Weather: Settler Vernaculars and Climate Anxieties in Early Twentieth-Century South Africa." *Environmental History* 23 (2018): 3–27.

McKittrick, Meredith. "Theories of 'Reprecipitation' and Climate Change in the Settler Colonial World." *History of Meteorology* 8 (2017): 74–94.

McLeod, Roy M. "Scientific Advice for British India: Imperial Perceptions and Administrative Goals, 1898–1923." *Modern Asian Studies* 9:3 (1975): 343–84.

Meh [née Lum], Choy Chee. "History of the Malaysian Branch of the Royal Asiatic Society." *JMBRAS, Sheppard Memorial Issue* 68:2 (1995): 81–148.

Menon, K. D. "A Brief History of Forest Research in Malaya." *Commonwealth Forestry Review* 48:4 (1969): 309–16.

Mercer, Harriet. "Atmospheric Archives: Gender and Climate Knowledge in Colonial Tasmania." *Environment and History* 27:2 (2021): 193–210.

Merritt, Brittany. "Insecurities of Empire: Struggles over Health Reform in Interwar Barbados." *Journal of Colonialism and Colonial History* 18:3 (2017): 1–13.

Mitchell, Timothy. *Rule of Experts: Egypt, Techno-Politics, Modernity*. Berkeley: University of California Press, 2002.

Monmonier, Mark. *Air Apparent: How Meteorologists Learned to Map, Predict, and Dramatize Weather*. Chicago: University of Chicago Press, 1999.

Moore, Peter. *The Weather Experiment: The Pioneers Who Sought to See the Future*. London: Penguin Random House, 2015.

Morgan, Ruth A. *Climate Change and International History: Negotiating Science, Global Change, and Environmental Justice*. London: Bloomsbury Academic Press, 2024.

Morgan, Ruth A. "Prophecy and Prediction: Forecasting Drought and Famine in British India and the Australian Colonies." *Global Environment* 13 (2020): 96–133.

Morgan, Ruth A. "Southern Skies: Australian Atmospheric Research and Global Climate Change." *Disaster Prevention and Management* 30:1 (2021): 47–63.

Mrázek, Rudolf. *Engineers of Happy Land: Technology and Nationalism in a Colony*. Princeton, NJ: Princeton University Press, 2002.

Murfett, Malcolm H., John N. Miksic, Brian P. Farrell, and Ming Shun Chiang. *Between Two Oceans: A Military History of Singapore from 1275 to 1971*. 2nd ed. Singapore: Marshall Cavendish Editions, 2011.

Musa, Mahani. "Early History: Penang before 1786." In *Early History of Penang*, edited by Muhammad Haji Salleh, 30–56. Penang: Penerbit USM, 2012.

Naylor, Simon. "Log-Books and the Law of Storms: Maritime Meteorology and the British Admiralty in the Nineteenth Century." *Isis* 106:4 (2015): 771–97.

Naylor, Simon. "Nationalising Provincial Weather: Meteorology in Nineteenth-Century Cornwall." *British Journal for the History of Science* 39:3 (2006): 407–33.

Naylor, Simon, and Matthew Goodman. "Atmospheric Empire: Historical Geographies of Meteorology at the Colonial Observatories." In *Weather, Climate, and the Geographical Imagination*, edited by Martin Mahony and Samuel Randalls, 25–42. Pittsburgh: University of Pittsburgh Press, 2020.

Nazarian, N., E. S. Krayenhoff, B. Bechtel, D. Hondula, R. Paolini, J. Vanos, T. Cheung, et al. "Integrated Assessment of Urban Overheating Impacts on

Human Life." *Earth's Future* 10:8 (2022): e2022EF002682. doi.org/10.1029/2022EF002682 (Open Access).

Nebeker, Frederik. *Calculating the Weather: Meteorology in the 20th Century.* California: Academic Press, 1995.

Ng, Beng Yeong. *Till the Break of Day: A History of Mental Health Services in Singapore, 1841–1993.* Singapore: NUS Press, 2016.

Ng, Francis S. P. *100 Years of Tropical Forest Research: The Story of the Forest Research Institute Malaysia.* Selangor: Forest Research Institute Malaysia, 2010.

Ng Siew Yoong. "The Chinese Protectorate in Singapore, 1877–1900." *Journal of Southeast Asian History* 2:1 (1961): 76–99.

Ngoei, Wen-Qing. *Arc of Containment: Britain, the United States, and Anticommunism in Southeast Asia.* Ithaca: Cornell University Press, 2019.

Nott, James. "Dance Halls: Towards an Architectural and Spatial History, c. 1918–65." *Architectural History* 61 (2018): 205–33.

Ophir, Adi, and Stephen Shapin. "The Place of Knowledge: A Methodological Survey." *Science in Context* 4 (1991): 3–21.

Paine, S. C. M. *The Wars for Asia, 1911–1949.* Cambridge: Cambridge University Press, 2012.

Pakiam, Geoffrey K. "'Not the Oil of the Country': Smallholders and British Malaya's Oil Palm Industry, 1929–1941." *Agricultural History* 95:1 (2021): 69–103.

Palladino, Paolo, and Michael Worboys. "Science and Imperialism." *Isis* 84 (1993): 91–102.

Pan Lu. *Image, Imagination and Imaginarium: Remapping World War II Monuments in Greater China.* Singapore: Palgrave Macmillan, 2021.

Parolini, Giuditta. "Building Networks of Exchange in Agricultural Meteorology: The Agro-meteorological Service in French Indochina." *History of Meteorology* 9 (2020): 1–20.

Patke, R., and P. Holden. *The Routledge Concise History of Southeast Asian Writing.* New York: Routledge, 2010.

Petitjean, Patrick, ed. *Colonial Sciences: Researchers and Institution.* Paris: Orstom Editions, 1996.

Phua Kai Hong and Mary Lai Lin Wong. "From Colonial Economy to Social Equity: History of Public Health in Malaysia." In *Public Health in Asia and the Pacific: Historical and Comparative Perspectives,* edited by Milton J. Lewis and Kerrie L. MacPherson, 170–87. London: Routledge, 2007.

Pickering, Andrew. "From Science as Knowledge to Science as Practice." In *Science as Practice and Culture,* edited by Andrew Pickering, 1–26. Chicago: Chicago University Press, 1992.

Pisani, Donald J. "Forests and Conservation, 1865–1890." *Journal of American History* 72:2 (1985): 340–59.

Powell, J. M. *Plains of Promise, Rivers of Destiny: Water Management and the Development of Queensland, 1824–1990.* Trove: Brisbane, 1991.

Pretel, David. "Hidden Connections: The Global History of Jungle Commodities." *Technology and Culture* 64:1 (2023): 202–19.

Pwee, Timothy. "Let There Be Light." *BiblioAsia* 16:4 (2021): 36–41.

Pyenson, Lewis. *Empire of Reason: Exact Sciences in Indonesia*. Leiden: Brill, 1989.

Quinn, W. H., and V. T. Neal. "The Historical Record of El Nino Events." In *Climate since A.D. 1500*, edited by R. S. Bradley and P. D. Jones, 623–48. Rev. ed. London: Routledge, 1995.

Radkau, Joachim. *Nature and Power: A Global History of the Environment*. Translated by Thomas Dunlap. Washington, DC: German Historical Institute, 2008.

Raj, Kapil. *Relocating Modern Science: Circulation and the Construction in South Asia and Europe, 1650–1900*. Basingstoke: Palgrave Macmillan, 2007.

Renn, Jürgen. *The Evolution of Knowledge: Rethinking Science for the Anthropocene*. Princeton, NJ: Princeton University Press, 2020.

Robertson, J. R. *The Battle of Penang: World War One in the Far East*. Kuala Lumpur: Trafalgar Publishing House, 2012.

Roth, Matthias, and Winston T. L. Chow. "A Historical Review and Assessment of Urban Heat Island Research in Singapore." *Singapore Journal of Tropical Geography* 33 (2012): 381–97.

Rouphail, Robert M. "Cyclonic Ecology: Sugar, Cyclone Science, and the Limits of Empire in Mauritius and the Indian Ocean World, 1870s–1930s." *Isis* 110:1 (2019): 48–67.

Sadka, Emily. "The Journal of Sir Hugh Low, Perak, 1877." *JMBRAS* 27:4 (1954): 5–108.

Salwi, Dilip M. "Madras Observatory: A Forgotten Page in Astronomy." *Journal of the British Astronomical Association* 98:4 (1988): 189–93.

Santamouris, M. "Using Cool Pavements as a Mitigation Strategy to Fight Urban Heat Island—A Review of the Actual Developments." *Renewable and Sustainable Energy Reviews* 26 (2013): 224–40.

Savage, Victor. *Western Impressions of Nature and Landscape in Southeast Asia*. Singapore: NUS Press, 1984.

Saw Swee-Hock. "Population Trends in Singapore, 1819–1967." *Journal of Southeast Asian History* 10:1 (1969): 36–49.

Schaffer, Simon. "Easily Cracked: Scientific Instruments in States of Disrepair." *Isis* 102:4 (2011): 601–835.

Schiebinger, Londa. *Plants and Empire: Colonial Bioprospecting in the Atlantic World*. Cambridge, MA: Harvard University Press, 2004.

Secord, James A. "Knowledge in Transit." *Isis* 95 (2004): 654–72.

Seth, Suman. "Putting Knowledge in Its Place: Science, Colonialism, and the Postcolonial." *Postcolonial Studies* 12:4 (2009): 373–88.

Shapin, Steven, and Simon Schaffer. *Leviathan and the Air-Pump: Hobbes, Boyle, and the Experimental Life*. Princeton, NJ: Princeton University Press, 1985.

Shennan, Margaret. *Out in the Midday Sun: The British in Malaya, 1880–1960*. Singapore: Monsoon Books, 2015.

Sheppard, Tan Sri Mubin. "Traditional Malay House Forms in Trengganu and Kelantan." *JMBRAS* 42:2 (1969): 1–9.

Simpson, Thomas. "Imperial Slippages: Encountering and Understanding Ice in Colonial India." In *Ice Humanities: Living, Thinking, and Working in a Melting World*, edited by Klaus Dodds and Sverka Sörlin. Manchester: Manchester University Press, 2022, 205–27.

Simpson, W. J. *Report on the Sanitary Condition of Singapore*. London: Waterlow, 1907.

Singh, D., R. Seager, B. Cook, M. Cane, M. Ting, E. Cook, and M. Davis. "Climate and the Global Famine of 1876–78." *Journal of Climate* 31 (2018): 9445–67.

Singh, Satyajit. *Taming the Waters: The Political Economy of Large Dams in India*. Delhi: Oxford University Press, 1997.

Siry, Joseph M. *Air-Conditioning in Modern American Architecture, 1890–1970*. University Park: Pennsylvania State University Press, 2022.

Smith, Philip. "Where There Is Life: Science Fiction in Singapore." *Journal of Commonwealth Literature* 57:2 (2019): 291–303.

Soojung-Kim Pang, Alex. *Empire and the Sun: Victorian Solar Eclipse Expeditions*. Stanford, CA: Stanford University Press, 2002.

Spicer, Chrystopher J. *The Flying Adventures of Jessie Keith "Chubbie" Miller: The Southern Hemisphere's First International Aviatrix*. Jefferson, NC: McFarland, 2017.

Strobl, Michael. "Alexander von Humboldt's Climatological Writings." *German Life and Letters* 74:3 (2021): 371–93.

Subrahmanyam, Sanjay. "Connected Histories: Notes towards a Reconfiguration of Early Modern Eurasia." *Modern Asian Studies* 31:3 (1997): 735–62.

Sutter, Paul S. "The Tropics: A Brief History of an Environmental Imaginary." In *The Oxford Handbook of Environmental History*, edited by Andrew C. Isenberg, 178–204. Oxford: Oxford University Press, 2014.

Takane, Yuya, Yukihiro Kikegawa, Masayuki Hara and C. Sue B. Grimmond. "Urban Warming and Future Air-Conditioning Use in an Asian Megacity: Importance of Positive Feedback." *npj Climate and Atmospheric Science* 2:39 (2019). doi.org/10.1038/s41612-019-0096-2/.

Takarabe, Kae. "The Smithsonian Meteorological Project and Hokkaido, Japan." *History of Meteorology* 9 (2020): 1–23.

Tang, Rahman A. "Revisiting the Water Issue across the Causeway: The Origins of Water Supply from Johor to Singapore, 1904–32." *Jebat: Malaysian Journal of History, Politics and Strategy* 40:1 (2013): 1–20.

Tan Yong Soon, Lee Tung Jean, and Karen Tan. *Clean, Green, and Blue: Singapore's Journey towards Environmental and Water Sustainability*. Singapore: ISEAS Publishing, 2009.

Tilley, Helen. *Africa as a Living Laboratory: Empire, Development, and the Problem of Scientific Knowledge, 1870–1950*. Chicago: Chicago University Press, 2011.

Tregonning, K. G. "Tertiary Education in Malaya: Policy and Practice, 1905–1962." *JMBRAS* 63:1 (1990): 1–14.

Turnbull, C. M. "The European Mercantile Community in Singapore, 1819–1867." *Journal of Southeast Asian History* 10:1 (1969): 12–35.

Turnbull, C. M. *The Straits Settlements, 1826–67: Indian Presidency to Crown Colony*. London: Athlone Press, 1972.

Turner, Roger. *Weathering Heights: The Emergence of Aeronautical Meteorology as an Infrastructural Science*. Publicly Available Penn Dissertations 147 (2010).

Tortajada, Cecilia, Yual Joshi, and Asit K. Biswas. *The Singapore Water Story: Sustainable Development in an Urban City-State*. London: Routledge, 2013.

Trombulak, Stephen C., ed. *So Great a Vision: The Conservation Writings of George Perkins Marsh*. Hanover: Middlebury College Press, 2001.

Udías, A. "Jesuits' Contribution to Meteorology." *Bulletin of the American Meteorological Society* 77:10 (1996): 2307–15.

van Bavel, Bas, Daniel R. Curtis, Jessica Dijkman, Matthew Hannafor, Maïka de Keyzer, Eline van Onacker, and Tim Soens, eds. *Disasters and History: The Vulnerability and Resilience of Past Societies*. Cambridge: Cambridge University Press, 2020.

Vogel, Brant. "The Letter from Dublin: Climate Change, Colonialism, and the Royal Society in the Seventeenth Century." *OSIRIS* 26 (2011): 111–28.

von Storch, Hans, and Nico Stehr. "Eduard Brückner's Ideas: Relevant in His Time and Today." In *Eduard Brückner: The Sources and Consequences of Climate Change and Climate Variability in Historical Times*, edited by N. Stehr and H. von Storch, 1–24. Dordrecht: Kluwer, 2000.

Walker, Malcolm. *History of the Meteorological Office*. Cambridge: Cambridge University Press, 2011.

Warren, James F. *Rickshaw Coolie: A People's History of Singapore, 1880–1940*. Singapore: NUS Press, 2003.

Warren, James F. "The Spiral of Failure: Suicide among the Singapore Rickshaw Coolies." *Southeast Asian Journal of Social Science* 13:2 (1985): 47–66.

Warren, James F. "Weather, History and Empire: The Typhoon Factor and the Manila Galleon Trade, 1565–1815." In *Anthony Reid and the Study of the Southeast Asian Past*, edited by Geoff Wade and Tana Li, 183–220. Singapore: ISEAS, 2012.

Weil, Benjamin. "The Rivers Come: Colonial Flood Control and Knowledge Systems in the Indus Basin, 1840s–1930s." *Environment and History* 12:1 (2006): 3–29.

Weiler, John. "Colonial Connections: Royal Engineers and Building Technology Transfer in the Nineteenth Century." *Construction History* 12 (1996): 3–18.

Werner, Anja. "Networks of Knowledge in the Sources of the 1826 Essai Politique sur l'Île de Cuba." *HiN - Alexander von Humboldt im Netz* 15:28 (2014): 59–67.

Westermann, Andrea. "Disciplining the Earth: Earthquake Observation in Switzerland and Germany at the Turn of the Nineteenth Century." *Environment and History* 17 (2011): 53–77.

Williams, Chris, ed. *A Companion to Nineteenth Century Britain: Blackwell Companions to British History.* 2nd ed. Maldon, MA: Blackwell Publishing, 2007.

Williamson, Fiona. "Building a Long-Time Series for Weather and Extreme Weather in the Straits Settlements: A Multi-disciplinary Approach to the Archives of Societies." *Climate of the Past* 17 (2021): 791–803.

Williamson, Fiona. "The Great Flood of 1926: Environmental Change and Disaster Governance in British Malaya." *Journal Ecosystem Health and Sustainability* 2:11 (2016). Special issue on the environmental impact of disasters.

Williamson, Fiona. "Just Doing Their Job: The Hidden Meteorologists of Colonial Hong Kong c. 1883–1914." *British Journal for the History of Science* 54:3 (2021): 341–59.

Williamson, Fiona. "Local Problems, Global Solutions? Making It Rain in Hong Kong c. 1890–1930." *History of Science* (2023). doi.org/10.1177/0073275322110./

Williamson, Fiona. "Malaya's 'Greatest Menace'? Slow Onset Disasters and the Politics of the Environment c. 1920–1950." *International Review of Environmental History* 4:2 (2018): 45–68.

Williamson, Fiona. "The Politics of Disaster: The Great Singapore Flood of 1954." *Environment and Planning E: Nature and Space* 1:3 (2018): 323–39.

Williamson, Fiona, and Clive Wilkinson. "Asian Extremes: Experience and Exchange in the Development of Meteorological Knowledge, c. 1840–1930." *History of Meteorology* 8 (2017): 159–78.

Wisner, Ben, Piers Blaikie, Terry Cannon, and Ian Davies, eds. *At Risk: Natural Hazards, People's Vulnerability and Disasters.* 2nd ed. London: Routledge, 2004.

Wong, K. M. "The Herbarium and Arboretum of the Forest Research Institute Malaysia at Kepong—a Historical Perspective." *Garden's Bulletin Singapore* 40:1 (1987): 15–30.

Wong Hong Suen. *Singapore through 19th Century Prints and Paintings.* Singapore: Editions Didier Millet and National Museum of Singapore, 2010.

Wong Yunn Chii and Tan Kar Lin. "Emergence of a Cosmopolitan Space for Culture and Consumption: The New World Amusement Park-Singapore (1923–70) in the Inter-War Years." *Inter-Asia Cultural Studies* 5:2 (2004): 279–304.

Yeoh, Brenda. *Contesting Space in Colonial Singapore: Power Relations and the Urban Built Environment.* Singapore: NUS Press, 2003.

Yeoh, Brenda. "Urban Sanitation, Health and Water Supply in Late Nineteenth and Early Twentieth Century Colonial Singapore." *Southeast Asia Research* 1:2 (1993): 143–72.

Zaiki, Masumi, and Togo Tsukahara. "Meteorology on the Southern Frontier of Japan's Empire: Ogasawara Kazuo at Taihoku Imperial University." *East Asian Science, Technology and Society: An International Journal* 1 (2007): 183–203.

Zilberstein, Anya. *A Temperate Empire: Making Climate Change in Early America.* New York: Oxford University Press, 2016.

Zuidervaart, Huib J., and Rob H. Van Gent. "'A Bare Outpost of Learned European Culture on the Edge of the Jungles of Java': Johan Maurits Mohr (1716–1775) and the Emergence of Instrumental and Institutional Science in Dutch Colonial Indonesia." *Isis* 95 (2004): 1–33.

INDEX

Note: Page numbers in *italics* indicate figures, and references following "n" refer notes.